DEPLOYING IPv6 IN BROADBAND ACCESS NETWORKS

DEPLOYING IPv6 IN BROADBAND ACCESS NETWORKS

Adeel Ahmed

Salman Asadullah

WILEY

A JOHN WILEY & SONS, INC., PUBLICATION

For general information on our other products and services or for technical support, please contact our Customer Care Department within the United States at (800) 762-2974, outside the United outside the United States at (317) 572-3993 or fax (317) 572-4002.

Wiley also publishes its books in a variety of electronic formats. Some content that appears in print may not be available in electronic formats. For more information about Wiley products, visit our web site at www.wiley.com.

Library of Congress Cataloging-in-Publication Data:

Ahmed, Adeel.
  Deploying IPv6 in broadband access networks/
  Adeel Ahmed and Salman Asadullah

Includes bibliographic references and index

ISBN 978-0-470-19338-9

10 9 8 7 6 5 4 3 2 1

I dedicate this book to my parents for everything they have done for me; I could never repay them. To my wife: for her love, support, and patience throughout the writing of this book. To my lovely children: Asad, Aashir, and Zeerak, you are my inspiration and my pride and joy. I love you all very much.

I would also like to dedicate the book to all the people suffering in this world and to those people who work endlessly to help others and to make the world a better place.

Adeel Ahmed

I dedicate this book to all who are suffering and going through tribulations, and to those who are working to relieve those in pain and suffering. I also dedicate this work to my loving parents, brothers, and two lights of my life: Bahira and Fatimah!

Salman Asadullah

CONTENTS

FOREWORD

The Internet is becoming a utility with an estimated 1.5 billion users, commonly referred to as "netizens," around the world. This large user base is surpassed only by the 3.5 billion mobile (cell) phone users on the planet. Approximately 10% of cell phones in use today are "smart phones," which also provide Internet services. The Internet backbone is quite robust; however, the last mile of the access layer is made up of a fragmented delivery system ranging from very low speed to high-speed (1 Gbps) connections. To put things in perspective, a 1-Gbps connection will allow downloading of a James Bond movie in about 20 seconds. High-speed download allows better use of Internet resources than does live streaming, which is resource intensive due to time constraints on packet delivery and reordering.

Broadband provides the Internet with an opportunity to become a robust utility similar to the TV networks. It is interesting to note that TV networks were designed with enough capacity to match viewer expectations for quality of experience. The next stage in the evolution of broadband access is to move from asymmetric to symmetric provisioning, thus allowing consumers to download and upload at the same speed. IPv6 provides a large address capacity and will be ideal for a commodity addressing scheme that will enable two-way always-on, Internet services. This scenario will signify the most important shift in adoption of the Internet, by empowering users to become "full-time residents" instead of simply sporadic in-and-out consumers. The evolving Internet offers a bright future by transforming casual users in TV broadcasters, reporters, and remote entertainers, and by creating new jobs and providing unprecedented opportunities that traditionally required expensive support infrastructure. The deployment of IPv6 has become an issue of strategic importance for many national economies. Telecom operators and Internet service providers (ISPs) are key players in facilitating the deployment of IPv6 on broadband access networks amid the obvious complexities of coexisting with or replacing widely deployed IPv4 services. Evidently, telecom operators and ISPs have to take steps to ensure a workable transition strategy that involves transparent interoperability and integration of mature and advanced applications over both IPv4 and IPv6. this strategy will enable a combination of services that will allow service providers to explore and exploit richer services offered by IPv6 during a potentially long transition from IPv4. This will also lead to new business models that will generate return on investment without waiting for the ubiquitous deployment of IPv6.

The IPv4 address space is virtually depleted, with just over 14% capacity remaining, and is expected to run out by the end of 2010. It is therefore anticipated that the use of IPv6 will gain momentum and the end user will drive differentiated services, achieving returns not only in investments but also in service innovations and flexible communication solutions. Solutions for integrating and deploying both IPv4 and IPv6 services are mature and available to service providers.

The authors of this book have the necessary technical expertise and experience to identify the challenges and to propose recommendations and solutions of great value to a world made of heterogeneous and widely un-interoperable networks designed using private addressing schemes that inhibit end-to-end applications and services. Their extensive involvement in standardization bodies such as the IETF and knowledge spread at the customer level in the design and deployment of IPv6 networks are of paramount value to readers, who can gain first-class knowledge to empower them to tackle the transition to IPv6 with greater confidence.

Welcome to the new two-way IPv6-based broadband access to the Internet!

LATIF LADID

President, IPv6 Forum

PREFACE

This book is meant to be used as a guide by network engineers and architects while deploying IPv6 in their broadband networks. Service providers worldwide are looking for ways to expand their networks and to meet the scalability requirements of the growing number of always-on devices connected to the Internet. IPv6 is currently the only solution available to meet these challenges and to enable service providers to scale their networks and provide new services to customers.

The focus of the book is on many of the challenges faced by service providers today and how IPv6 addresses these issues. Chapter 1 covers drivers for IPv6 in broadband networks to give readers an idea of why they should be looking at deploying IPv6 in their networks. Chapter 2 provides an overview of IPv6 protocol basics and provisioning.

Chapters 3 and 4 are concentrated on IPv6 deployment techniques in various broadband networks, such as cable, DSL, ETTH, and wireless. Comparisons are drawn between IPv4 and IPv6 deployment models, and similarities and differences between the two are discussed in detail.

The concluding four chapters deal in detail with the configuration of different network components in an IPv6 broadband solution, and provide guidelines on debugging and troubleshooting IPv6-related issues. This will help readers configure and troubleshoot problems when they deploy IPv6 in their networks.

Appendix A has a real-life SP case study of Free's IPv6 broadband deployment. Free is the second-largest French ISP with more than four million broadband subscribers (ADSL and FTTH).

This book is intended for those network engineers and architects who are contemplating deploying IPv6 in their broadband networks. The book contains detailed information as to how IPv6 can be deployed in service provider broadband networks. Different IPv6 deployment models, configurations, and troubleshooting guidelines are discussed to help readers understand the challenges faced by SP in deploying IPv6 in broadband access networks.

ADEEL AHMED
SALMAN ASADULLAH

ACKNOWLEDGMENTS

First and foremost, the authors acknowledge Benoit Lourdelet for his major contributions to Chapter 4 and for reviewing the remaining chapters and providing valuable feedback. The authors also thank Latif Ladid and Patrick Grossetete most sincerely for honoring us by endorsing the book and by writing the foreword and the endorsement notes.

The authors appreciate the hard work of all the reviewers, whose valuable insights and feedback improved the book significantly. Our special thanks go to Abe Martey for making available his vast knowledge of networking technology and for his rigorous review of the text. Abe's prior experience as the author of two networking technology books, *IS-IS Network Design Solutions* and *Troubleshooting IP Routing Protocols*, both published by Cisco Press, and his detailed comments and feedback, were great assets and made our work more valuable for the audience.

We would also like to thank all the people whom we have worked with on the IPv6 front at Cisco, vendors, operators, IETF, other standards bodies and platforms, and industry at large, to make IPv6 a reality.

We especially thank George Telecki, Michael Christian, Angioline Loredo, and all the staff at Wiley for helping us drive this book to completion.

Special Acknowledgments by Salman Asadullah: I would like to thank a few people who although not directly related to this work have helped me through the journey of life in one way or an other: Muhammad ibn 'Abdullah and his companions, Mohammad Asadullah, Shakila Siddiqui, Salah Uddin Ayubi, Salik bin Saddina, Hamaza Yousf, Sohaib Webb, Usama Canon, Anwar Awlaki, Abdul Sattar Edhi, Dr. Ali Metwally, Dr. Magda Mohsen, Imran Asadullah, Bahira Metwally, Kamal Siddiqui, Khalid Raza, Nasir Kamal, Romana Khan, Mike Quinn, Syed Khurram, John Selden, Iqbal Ahmed Khan, Abdul Mateen Hashmi, Adeel Ahmed, Fawad Asadullah, Zulfiqar Ahmed, Rasheed Uddin, Robert Santiago, and Himanshu Desai.

Special Acknowledgments by Adeel Ahmed: I would like to acknowledge a few people who have especially influenced my life and have provided guidance and inspiration: Muhammad ibn 'Abdullah and his companions, Sanjeeda Ahmed, Aziz Ahmed, Shakila Khanum, Khairunisa Begum, Shaikh Suhaib Webb, Shaikh Saad Hassanin, Natasha Ahmed, Najia Ahmed, Farzana Khan, Umar Saeed, Shaukat Khalil, Nasir Ali, and Salman Asadullah.

ABOUT THE AUTHORS

This book is written by Cisco Certified Internetwork Experts (CCIEs) who have been working with various customers worldwide on IPv6 deployment, standard bodies, and technical forums since 2002. The authors have worked with several development teams within Cisco in driving IPv6 implementation on Cisco products and influencing the IPv6 technology direction through their work with standards bodies. They have written and contributed to several white papers, design guides, and IETF RFCs and drafts on deploying IPv6, and have provided numerous trainings and seminars on this topic. The authors' combined 22 years of Internetworking industry experience and 12 years of experience in working with IPv6 brings valuable knowledge and expertise to the book.

Adeel Ahmed, CCIE No. 4574, is a Technical Leader in Cisco's Advanced Services group. He has been with Cisco Systems for over 10 years. His areas of expertise include access/dial, broadband cable, and IPv6. He has worked with major cable MSOs in North America, EMEA, and ASIAPAC in designing and troubleshooting cable networks. He has written several white papers and design guides used by customers, sales teams, and Cisco engineers in deploying multiservices over cable networks. He has also coauthored and contributed to IETF RFCs and drafts.

Ahmed has represented Cisco at industry technical forums such as IETF, CableLabs, NCTA, SCTE, Networkers, APRICOT, NAv6TF, and Global IPv6 Summit. He holds a bachelor's and a master's degree in electrical engineering.

Salman Asadullah, CCIE No. 2240, is a Technical Leader at Cisco Systems and holds honorary positions at APRICOT, NSP, and IPv6 Forum. Recognized as an expert within Cisco and by Cisco's customers and industry as a whole, he has been designing and troubleshooting large-scale IP and multiservice networks for over 13 years. He has represented Cisco in industry panel discussions and technical platforms such as Networkers, APRICOT, NANOG, SANOG, IETF, and IPv6 Forum events.

Asadullah influences technology and product directions and decisions within Cisco business units and in the Internet community. He has produced several technical documents, white papers, and articles and has coauthored and contributed to IETF RFCs and drafts. He is a coauthor of two networking technology books: *Cisco CCIE Fundamentals: Network Design & Case Study,* and *PDIO of the IP Telephony Networks*, both published by Cisco Press. *PDIO of the IP Telephony Networks* is a best-seller, with over 13,000 copies sold to date. He holds a B.S. in electrical engineering from Arizona and a M.S. in electrical engineering from Kansas.

ABOUT THE CONTRIBUTORS

Benoit Lourdelet is an IPv6 Senior Product Manager at Cisco Systems. He helps to drive the range of Cisco products that support IPv6. Benoit has been a key player in the deployment and architecture of the first IPv6 broadband networks across the globe. As a known IPv6 expert, he continues to influence the architecture designs of next-generation IPv6 broadband networks. He has over 15 years of experience in designing and operating telecommunications networks. He has also worked with both manufacturers and service providers in developing Internet exchange points such as PARIX and international backbones. He has contributed toward many IPv6 technical papers and IETF RFCs and drafts. He is regular speaker on IPv6 topics both at Cisco events and at IPv6 international conferences such as Cisco Networkers, AFNOG, and IPv6 Forum events. He holds an M.S. in chemistry and an M.S. in computer science from institutions in France.

Alexandre Cassen works as a software architect at Freebox, the R&D lab of Free.fr. He focuses on large-scale protocol and software design and implementation of value-added services. Most of his time is dedicated to developing software for IPTV and VoIP networking components. He enjoys learning new networking protocols and technologies.

ABOUT THE REVIEWERS

Abe Martey, CCIE No. 2373, works in the focused technical support group in Cisco Systems, where he provides expert support on Cisco's high-speed router platforms and other SP core networking technologies to major worldwide SPs. Abe is the author of *IS-IS Network Design Solutions* and coauthor of *Troubleshooting IP Routing Protocols*, both published by Cisco Press. Abe has been with Cisco for over 10 years, during which he has held several positions ranging from technical support and technical and product marketing. Prior to Cisco, Abe was a network engineer at Sprint, where he worked in Sprint's Managed Router Network Group and also as a support engineer in the early days of the Sprintlink IP Network. Abe has an MS in electrical engineering and is an active member of the IEEE.

Srinivasa Neppalli, CCIE No. 6370, works as a Network Consulting Engineer in the Advanced Services Broadband Team at Cisco Systems. He has been with Cisco since 1999. He holds CCIE certifications in R&S and service provider tracks. He has been providing consulting service for service providers and cable MSOs for the last seven years. He focuses on IP backbone technologies and his expertise includes routing protocols, MPLS applications, multicast, and IPv6. He holds a master's degree in electrical engineering.

Roy Boos is a Network Consulting Engineer in the Cisco Advanced Services Group. He has been with Cisco for over 10 years. His areas of expertise include network management, SNMP, broadband cable, and provisioning. His customers have included major MSOs in North America and ASIAPAC. He has provided detailed audits of their provisioning systems and authored several best practices white papers on monitoring recommendations for the 7246VXR, uBR10K, and 7600 via SNMP. Roy holds a bachelor's degree in computer engineering.

Michael Reekie is a Network Consulting Engineer in the Cisco Systems Advanced Services (AS) team. Michael has been involved in the testing and deployment of the Cisco Network Registrar (CNR) DHCP and DNS servers, and the Broadband Access Center family of products since joining Cisco in 1999 as part of the American Internet acquisition. In 2001, he played a key role in performance testing of the redesigned CNR DHCP protocol engine. Later he created an in-depth training program for engineers tasked with supporting CNR in the Cisco Technical Assistance Center (TAC). Michael joined Cisco Advanced Services in 2006 and has been supporting Tier 1 Cable MSO customer' provisioning data and voice services since then. He received his bachelor's degree in 1993 from Boston University.

1 IPv6 Drivers in Broadband Networks

With the exponential growth of the Internet and an increasing number of end users, service providers (SPs) are looking for new ways to evolve their current network architecture to meet the needs of Internet-ready appliances, new applications, and new services. Internet Protocol Version 6 (IPv6) is designed to enable SPs to meet these challenges and provide new services to their customers.

The life of IPv4 was extended by using techniques such as network address translation (NAT) and other innovative address allocation schemes. However, the need for intermediate nodes to manipulate data payload while employing these schemes posed a challenge to peer-to-peer communications, end-to-end security, and quality of service (QoS) deployments. IPv6 also addresses fundamental limitations in the IPv4 protocol that renders the latter incapable of meeting long-term requirements of commercial applications. Besides its inherent capabilities to overcome the aforementioned limitations, IPv6 also supports an address space quadruple of that of IPv4, by supporting 128-bit instead of 32-bit addresses (RFC3513). The huge IPv6 address space will enable IPv6 to accommodate the impending worldwide explosion in Internet use. IPv6 addressing provides ample addresses for connecting consumer home/Internet appliances, IP phones for voice and video, mobile phones, web servers, and so on, to the Internet without using IP address conversion, pooling, and temporary allocation techniques. IPv6 is designed to enhance end-to-end security, mobile communications, and QoS, and also to ease system management burdens, as the protocol is still evolving, with some of its capabilities still a work in progress by the Internet Engineering Task Force (IETF).

As the number of broadband users increase exponentially worldwide, cable, digital subscriber line (DSL), Ethernet to the home (ETTH), wireless, and other always-on access technologies, along with IPv6's huge address range, long-lived connections to servers, and permanent prefixes for home appliances, give more flexibility to SPs. Specifically, the addressing capacity of IPv6 has made it valuable to SPs, most of which are rolling out IPv6 support in their networks or aggressively evaluating its potential and value for service delivery. Outside the

Deploying IPv6 in Broadband Access Networks, By Adeel Ahmed and Salman Asadullah
Copyright © 2009 John Wiley & Sons, Inc.

United States, IPv6 adoption is being promoted on a national level, and countries such as Japan, Korea, China, India, and some European countries have taken lead roles in moving from testing and evaluation to actual deployment in broadband services and applications.

Cable, DSL, ETTH, and wireless services are the main broadband technologies that are widely deployed. In this book we discuss key aspects of IPv6-enabled broadband networks and explore differences from IPv4 deployments.

1.1 IPv6-BASED SERVICES

Until recently, IPv6-based services were considered primarily as differentiators that allowed SPs to exploit the large address space available in IPv6 for future growth planning and as a competitive advantage. However, as IPv6 has become more popular and familiar, SPs are adopting the protocol not only to offer new services to their customers but also for provisioning and managing a large number of network devices and applications. Governmental interest and promotion by means of incentives and favorable legislation is also a major driver for the growing adoption of IPv6 in SP and enterprise networks.

SPs in densely populated regions such as Asia and Europe are at the forefront of adoption and integration of IPv6 into their networks to address the increasing numbers of broadband subscribers and the greater scaling of their networks. For example, Nippon Telephone and Telegraph (NTT) in Japan is currently offering dual-stack commercial services to asymmetric digital subscriber line (ADSL) and fiber to the home (FTTH) subscribers. Dual-stack devices are capable of forwarding both IPv4 and IPv6 packets; hence, these FTTH and ADSL users can access the services with either an IPv4 or an IPv6 address, or both. In this deployment model, subscribers are offered /64 dedicated IPv6 prefixes but generally receive only a single static or dynamic IPv4 address.

Additionally, some SPs are offering integrated IPv6-based multicast and voice over IP (VoIP) service in addition to existing IPv4-based services, and are taking advantage of the larger IPv6 address space and other useful features. The multicast services consist of several video and audio streams available simultaneously to broadband subscribers. The content providers store the content based on users' interests and send this content as multicast streams to broadband subscribers. Today, with IPv4 service offerings, generally a single device attached directly to a gateway router (GWR) at the customer's premises receives the multicast stream. In similar IPv6 offerings, multiple devices may be attached to the GWR, each receiving a different stream at the same time.

For example, in Japan cable and satellite TV are not very popular, and users expect to receive video content through traditional broadcast TV programs. This provides an opportunity for content service providers to generate additional revenue by offering content not available through TV to broadband subscribers at reasonable prices. The content provider may multicast several channels of video and audio, and broadband subscribers will join various multicast groups of interest to receive content. Disney movies are an example of a video stream, and an audio stream could be karaoke. In this regard, this service offering is similar to a cable TV subscription.

In North American and the Asia Pacific region, IPv6 is being adopted primarily by large cable multiple systems operators (MSOs) to address growing subscriber-base and IP-enabled devices. Currently, cable MSOs use RFC1918 address space for cable modems and set-top-box management. RFC1918 provides 16 million addresses under the 10.0.0.0/8 prefix, plus 1 million addresses under 172.16.0.0/12 and 65,000 addresses under 192.168.0.0/16 prefixes. Factually, address utilization efficiency decreases with hierarchical topologies (see the HD ratio in RFC1715 and RFC3194), so the 9.8 million cable modems and set-top boxes could easily exhaust all 16 million RFC1918 private addresses. The exhaustion of IPv4 private address space among the cable MSO community has become a catalyst in the definition and standardization of the data over the cable service interface specification (DOCSIS) 3.0 standard, which introduces support of IPv6 in cable networks. Now, cable MSOs are planning to deploy IPv6 to manage this large number of cable modems and set-top boxes.

SPs are also offering IPv6 services over wireless links using 802.11-compliant WiFi hot spots. This enables users to take notebook PCs and PDAs along with them and connect to the Internet from various locations. One of the potential benefits of this service flexibility may be downloading digital pictures from a mobile phone with a digital camera to a home storage server.

Figure 1.1 depicts an end-to-end SP network with some of the most commonly deployed access broadband technologies, a core network and back-end provisioning, and network management servers. The access layer of the network features different terminologies for provider edge (PE) devices in the various access models; however, in all cases, PE devices function similarly by acting as head-end devices. The head-end devices connect to multiple downstream customer premises devices using different encapsulation techniques, protocols, and methodologies. In this book we explore the integration and deployment of IPv6 in the broadband access segment as well as in provisioning and network management servers. We also highlight some commonly used techniques for enabling the network core for carrying IPv6 traffic. Once the SP has enabled IPv6 in an access broadband, core, and on back-end provisioning and management servers, it may consider enabling IPv6 on a per application basis. For example, the SP may choose to manage the network devices using IPv6 as a first application.

1.2 BROADBAND ACCESS MODELS

With the exception of cable MSOs, two access models are prevalent in most access broadband technologies, such as ETTH, DSL, and wireless local area

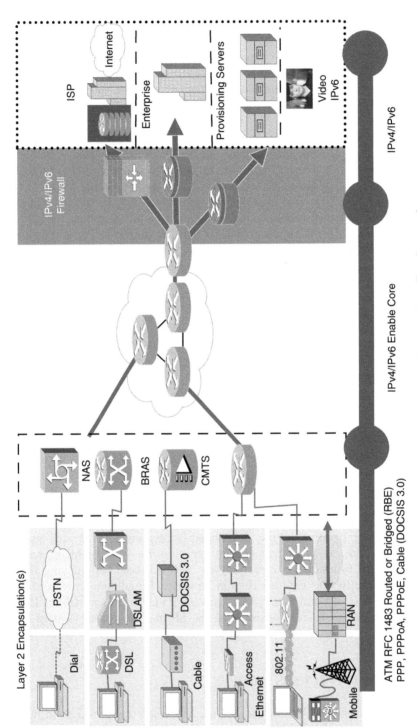

FIGURE 1.1 Typical end-to-end SP access broadband network.

4

networks (WLANs). In the first model the network access provider (NAP) and the network service provider (NSP) are owned and operated by the same entity. This is referred to as the ISP-operated model. In the second model the NAP and the NSP are owned and operated by two separate entities. This is known as the wholesale deployment model.

Generally, NAP is the entity that provides last-mile access services, and NSP is the entity that provides Layer3 services to customers. NAP and NSP operations are explained in detail in Chapter 4. Cable MSO architectures are very different from ETTH, DSL, and WLAN architectures and are covered in detail in Chapter 3.

1.2.1 ISP-Operated Deployment Model

In an ISP-operated model, a PE router [also known as an edge router (ER)], which is owned by the broadband SP, assigns the IPv4 address to the GWR. This deployment model is depicted in Figure 1.2. Assignment of the IPv4 address is done either by DHCPv4 or by static configuration. The IPv4 traffic is sent from the GWR to the PE using point-to-point protocol over Ethernet (PPPoE) (RFC2516), point-to-point protocol over ATM (PPPoA) (RFC2364), or routed bridged encapsulation (RBE) access methods, which are valid encapsulation candidates to offer IPv6 connectivity in a variety of access SP designs. Methods that are PPP based can leverage the IPv6 extensions to the authentication authorization and accounting (AAA) framework and the remote authentication dial-in user service (RADIUS) protocol (RFC3162) and fit in well with current IPv4 deployment models.

When deploying IPv6 in this framework, the same encapsulation techniques are used and the same behavior is achieved by ER assigning the IPv6 address to the GWR. The assignment of an IPv6 address is done either by using stateless address autoconfiguration (SLAAC), DHCPv6 using AAA and RADIUS, or static configuration. Depending on the agreement between the broadband SP and the user, this IPv6 address assignment could have a prefix length of /64 or shorter. The ER and GWR are upgraded to dual-stack routers in order to support both IPv4 and IPv6. The only scenario when GWR does not need to be

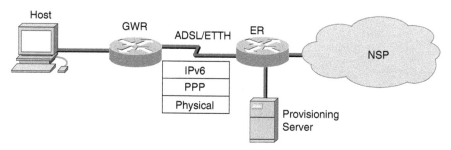

FIGURE 1.2 ISP-operated deployment model.

in dual-stack mode is when the host behind the GWR is running a PPPoE client and the IPv6 address is assigned directly to the host by the ER router, with the GWR acting only as a bridge.

1.2.2 Wholesale Deployment Model

In the wholesale deployment model the NAP only provides network access to users; it does not deal with addressing issues. NAP transports the user's IPv4 traffic from the GWR to the NSP ER [which acts as an L2TP network server (LNS)] by various means, such as Layer2 circuits [virtual local area networks (VLANs)], ATM permanent virtual circuits (PVCs), or frame relay virtual circuits (VCs), or by using other common encapsulation techniques (these PPP sessions can further be bundled be into an L2TP tunnel). The NAP's L2TP access concentrator (LAC) initiates a L2TP tunnel to the NSP's ER router. The LNS assigns IPv4 addresses to the devices (routers, PCs, etc.) located behind the GWR using DHCPv4 with AAA and RADIUS or by static configuration. Once the L2TP tunnel comes up, all traffic is forwarded to the NSP's ER over the L2TP tunnel. In a nutshell, the NSP's ER terminates these circuits and acts as a Layer3 gateway for all these users. For this reason, the NSP is actually responsible for assigning IP addresses to users. Figure 1.3 illustrates the wholesale deployment model operation.

In this model, if the NSP wants to provide content via multicast, it has to replicate that content to all subscribers. To preserve network resource one tries to do the replication as close to users as possible. In the case of an NSP reaching its customers through a wholesale NAP, the closest Layer3 device to the users is the NSP ER that terminates the virtual circuits. This means that the NSP will have to replicate the packets for all the virtual circuits with the users who requested it, and flood the NAP infrastructure with multicast replications. This is not a problem for the NSP; however, the wholesale NAP now has its network flooded with duplicate packets. This is not optimal, and it limits dramatically the capability of the wholesale NAP to scale support for the NSP's IP multicast service. Chapter 4 covers in detail new deployment models supported by IPv6 for addressing such scaling challenges.

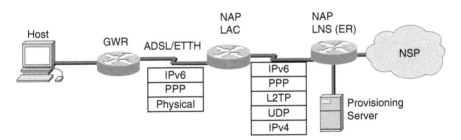

FIGURE 1.3 Wholesale deployment model.

When integrating IPv6 support in this model, the GWR and NAP's LAC are upgraded to dual-stack routers to support both IPv4 and IPv6. It is not necessary to upgrade the GWR to dual-stack capability when the host behind the GWR is running a PPPoE client, and the IPv6 address is assigned directly to the host from the NSP ER, with GWR acting only as a bridge. The NAP's LAC initiates a L2TP tunnel to the NSP's ER to forward traffic received from the GWR. The NSP ER assigns IPv6 addresses to the devices (routers, PCs, etc.) located behind the GWR using SLAAC, DHCPv6 using AAA and RADIUS, or static configuration. The GWR may receive a shorter than /64 prefix, depending on the NSP's policy and the customer's requirements. In this case the NSP's ER router is also upgraded to dual-stack status to support both IPv4 and IPv6.

1.3 SUMMARY

IPv6 enables SPs to offer new services as well as to enhance current services with a focus on servicing endpoints. IPv6 services may range from network addressing support for home appliances to peer-to-peer communication, such as Internet gaming, music and video file sharing, and IP telephony. With sufficient address space in the foreseeable future for several billion subscribers, appliances, and applications, IPv6 is the gateway to the future of the next-generation Internet. It is interesting to note that the world's population is projected to be 10 billion in 2050 and that up to 34 billion IPv6 addresses will be available to be assigned to each person as well as to many of the planets appliances, automobiles, buildings, cameras, control units, embedded systems, home networks, medical devices, mobile devices, monitors, output devices, phones, robots, sensors, switches, and VPNs. Thus, IPv6 holds the key to the success of the next-generation Internet.

REFERENCES

1. S. Asadullah and A. Ahmed, "IPv6 in Broadband," Cisco Systems, Inc. Packet Magazine, Fourth Quarter 2004.
2. S. Asadullah, A. Ahmed, C. Popoviciu, P. Savola, and J. Palet, "ISP IPv6 Deployment Scenarios in Broadband Access Networks," RFC4779, January 2007.
3. B. Lourdelet, "Application Note: IPv6 Access Services," Cisco Systems, Inc.
4. B. Lourdelet, "Application Note: DHCPv6," Cisco Systems, Inc.
5. Cisco Systems Inc., tutorial on IPv6 basics: "The ABC of IPv6."

2 IPv6 Overview

The IPv6 protocol evolved from years of operational experiences with the IPv4 protocol. In defining the IPv6 protocol, the networking industry and the IETF community tried to overcome real and perceived shortcomings of IPv4. A number of IETF RFCs cover currently defined capabilities of IPv6. Each RFC was defined after rigorous and long debate in the IETF. In this chapter we highlight some of the basics of IPv6 protocol, such as operations, addressing, and provisioning, all of which are important for engineers and architects who plan, design, and deploy IPv6 in access broadband networks. In this chapter we set the stage for technical deployment issues discussed in later chapters.

2.1 IPv6 PROTOCOL BASICS

2.1.1 IPv4 and IPv6 Header Comparison

In this section we briefly review the differences and similarities between the IPv4 and IPv6 headers and discuss the rationale behind the changes introduced into IPv6. IPv6 has a fixed 40-byte header encompassing eight fields. A quick look at Figure 2.1 shows that the IPv6 header inherits some fields from IPv4. Also, some fields are simply renamed, whereas others have been eliminated and new one(s) added.

The following fields in the IPv6 header are passed on from IPv4, including their original functions.

- *Version* This is a 4-bit field that contains (0110) to indicate IPv6 instead of (0100), which was used for the IPv4 version.
- *Source Address* A 32-bit field for IPv4, this is now a 128-bit field, to represent the larger IPv6 source address.
- *Destination Address* This is a 128-bit field to represent an IPv6 destination address. This was a 32-bit field for IPv4.

The following fields in the IPv6 header are renamed to better describe them while retaining their functionalities as inherited from IPv4.

Deploying IPv6 in Broadband Access Networks, By Adeel Ahmed and Salman Asadullah
Copyright © 2009 John Wiley & Sons, Inc.

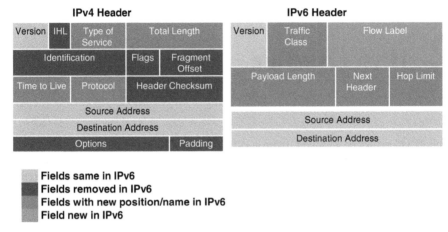

FIGURE 2.1 IPv4 and IPv6 header comparison.

- *Traffic Class* an 8-bit field similar to the type of service (TOS) field in IPv4. It tags the packet with a traffic class that can be used in differentiated services.
- *Payload Length* similar to the total length field in IPv4, except that it does not include the 40-byte header, since the IPv6 header is fixed at 40 bytes.
- *Hop Limit* similar to the time to live (TTL) field in IPv4; decrements by one at each router.
- *Next Header* similar to the protocol field in IPv4. The value in this field specifies what type of information follows the 40-byte IPv6 header. For example, it could be the transmission control protocol (TCP), user datagram protocol (UDP), internet message control protocol version 6 (ICMPv6), or an extension header (EH).

The following fields present in the IPv4 header are not available in the IPv6 header.

- *Header Length Field (IHL)* IHL was used to define the length of the header since the IPv4 header is of variable length. IPv6 has a fixed header length (40 byte), so there is no longer any need for this field.
- *Fragmentation Field* The fragmentation field is used to support packet fragmentation in order to accommodate transmission over interfaces that support smaller packet sizes. Routers do not fragment IPv6 packets [rather, the end host is responsible for IPv6 packet fragmentation after using path maximum transmission unit discovery (PMTUD)], due to the associated inefficiencies; that is, loss of one fragment of a packet generally causes complete retransmission of the entire set of fragments of the packet. Instead, the originating node or source node of a packet uses the PMTUD

process to determine the maximum MTU in a path between the source and the destination before sending the data. Since PMTUD is based on ICMPv6, its it has little effect on bandwidth and processing use.

- *Identification Field* The identification field is used in conjunction with the source IP address to uniquely identify a datagram as it leaves the source. This is helpful in reassembling packets from fragments. Since there is no fragmentation in IPv6, there is no need for the identification and flag fields.

- *Header Checksum Field* In IPv4 the checksum is done at the media access layer, network layer, and transport layer; this is considered to be excessive considering that IP transmission is a best-effort mechanism. Therefore, the checksum field is removed from the IPv6 header, expediting packet processing. IPv6 relies on the media access layer and transport layer (TCP or UDP) checksums. When IPv6 application uses TCP, checksum verification is automatic. However, for UDP applications, checksum verification is not automatic and UDP-based IPv6 application is required to enable the UDP checksum.

A new 20-bit flow label field is added to the IPv6 header. At the time of writing this book, definite applications and use of the flow label are still being discussed. However, RFC3697 provides a general definition, and the application of the flow label is that each source chooses its own flow label values; routers use a source address and flow label to identify distinct flows. A possible future application of this field could be to prioritize flows of varying importance.

2.1.2   IPv6 Extension Headers

The Next Header field indicates the type of information following the 40-byte IPv6 header. This could be TCP, UDP, or ICMPv6 data or an extension header (EH). Since IPv6 has a fixed 40-byte header, and if there are specific processing needs for a certain IPv6 packet, the header may be augmented by using EHs. The EH field is optional, and multiple EHs (a chain of EHs, Figure 2.2) could be used on a single IPv6 packet as needed. Each EH is identified by the Next Header field of the preceding header. The final EH field in the chain will have a Next Header field pointing to a transport-layer protocol such as TCP, UDP, or ICMPv6.

There are many types of EHs. Each EH should occur at most once, except for the Destination Options header, which should occur at most twice. As mentioned earlier, multiple EHs could be used in a single IPv6 packet. An example is a situation where encryption and authentication need to be performed on a packet, so an authentication header (51) and an ESP header (50) are used in parallel. Having more EHs tied to a packet means more processing resources on devices.

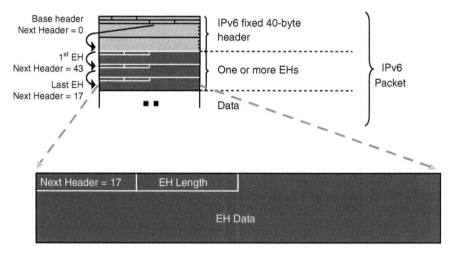

FIGURE 2.2   Chain of IPv6 extension headers.

When multiple EHs are used in the same packet, the order of the headers follows certain rules. For example, if multiple EHs are used and a Hop-by-Hop EH is present, it has to be on the top, because this packet needs to be processed by each hop along the way and should not buried under other EHs. A Hop-by-Hop EH is used for applications such as RSVP, where bandwidth has to be reserved at each hop along the path from source to destination, being processed at every hop along the way. The IPv6 EHs are documented in RFC2460.

Some of the best known EHs are:

Hop-by-Hop header   0
Destination Options header (with a routing header)   60
Routing header   43
Fragment header   44
Authentication header   51
ESP header   50
Mobility header   135
Destination Options header   60
ICMPv6   58
No Next header   59

2.1.3 IPv6 Addressing

The key driver for IPv6 was to make available more IP addresses for the ever-expanding Internet, with a significantly growing number of connected devices. The IPv6 address is 128 bits long rather than the 32 bits provided in IPv4. This allows for a significantly larger address space in IPv6, eliminating the need to share addresses from a pool or to use network address translation (NAT), both austerity measures used to conserve IPv4 addresses from near-term depletion. There is a fundamental difference between IPv4 and IPv6 addressing architecture. The details are captured in RFC3513. However, in this section we highlight some key differences.

The 128-bit IPv6 address is broken down into eight chunks of 16-bit hexadecimal numbers separated by colons. These alphanumeric hexadecimal numbers are not case sensitive. An example of a 128-bit IPv6 address is

```
2001:0db8:0000:130F:0000:0000:087C:140B
```

In this example, each hex digit represents 4 bits, so 2001 is equivalent to 0010 0000 0000 0001, giving a total of 16 bits in each of the eight chunks above.

Another important thing to remember is that these long 128-bit addresses can be shortened by compressing the leading zeros of any of the chunks and by representing zeros in a contiguous block by a double colon (::). Double colons can appear only once in an IPv6 address; otherwise, we could not determine the number of zero bits represented by each double colon. The compressed version of the earlier example is represented as

```
2001:db8:0:130F::87C:140B
```

The IPv6 prefixes can be represented as prefixes with length indications, similar to the manner in which it is done in IPv4. In IPv4 a class A address is represented as 10.0.0.0/8. In IPv6 an address is represented as 2001:0db8:0012:0000:0000:0000:0000:0000/48 but could be compressed and represented as 2001:db8:12::/48. Here the leading zeros are suppressed for two 16-bit chunks and six 16-bit chunks of all-zeros are represented as a double colon. Here we also keep in mind that the IPv6 addressing architecture does not present a concept of IPv4 address classes as used in classless interdomain routing (CIDR).

IPv6 address types are divided into four categories:

1. *Unicast*: one-to-one (global, link-local, unique-local, loopback, unspecified, IPv4 compatible, IPv4 mapped)
2. *Anycast*: one-to-nearest
3. *Multicast*: one-to-many
4. *Reserved*

A single interface may be assigned multiple IPv6 addresses of any type (unicast, anycast, multicast). The first few bits of any of these addresses tell us what type of address we are dealing with. There is no more broadcast address in IPv6; rather, multicast address is used to achieve the same functionality.

2.1.3.1   Unicast Address

Global Unicast Address   As mentioned earlier, there are several types of unicast addresses. The most important is the global unicast address (also called the global unicast IPv6 address), which is the equivalent of the IPv4 global unicast address. A global unicast address is an IPv6 address from the IPv6 global unicast address range. The definition of the IPv6 address space enables strict aggregation of routing prefixes to limit the number of routing table entries in the global routing table. Global unicast IPv6 addresses are aggregated by organizations and advertised to their respective Internet service providers (ISPs). The ISPs further aggregate these prefixes and advertise them in a global IPv6 routing table. 2000::/3, the global unicast address range, uses one-eighth of the total IPv6 address space. It is the largest block of assigned IPv6 addresses today.

As shown in Figure 2.3, the IPv6 address has two major parts, the network prefix and the interface ID. The global unicast address typically consists of a 48-bit global routing prefix and a 16-bit subnet ID. The 16-bit subnet field could be used by individual organizations to create their own local addressing hierarchy and to identify up to 65,535 individual subnets. The interface ID is the lowest-order 64-bit field of unicast address used to identify a unique interface on a link. A link is a network medium over which network nodes communicate using the link layer. This interface ID may be assigned in several different ways:

1. Autoconfigured from a 64-bit extended unique identifier (EUI-64)
2. Autogenerated pseudorandom number (for privacy concerns)
3. Assigned via DHCP
4. Manually configured

For hosts, the most commonly used is the EUI-64 format interface ID, which is derived from the 48-bit link-layer (MAC) address by inserting the hex

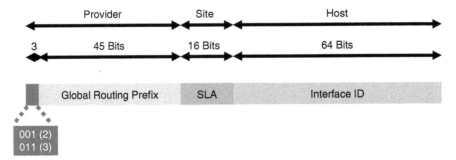

FIGURE 2.3   Structure of a global unicast address.

number FFFE between the upper three bytes of the organizationally unique identifier (OUI) field and the lower three bytes (serial number) of the link-layer address. To make sure that the address chosen is from a unique Ethernet MAC address, the seventh bit in the high-order byte is set to 1 (equivalent to the IEEE G/L bit) to indicate the uniqueness of the 48-bit address:

`2001:DB8:0000:0000:212:7F`**`FF:FE`**`EB:6A50`

The autogenerated pseudorandom number is a new concept, used to address privacy concerns. This allows the interface ID to be generated randomly and also has the ability to regenerate from time to time. The idea behind this approach is to decouple the MAC address and interface ID to avoid associating the source IPv6 address with a particular network adapter. It is next to impossible for a hacker to trace an IPv6 address using a randomly generated interface ID. This is due to the huge number of possible combinations of interface IDs (2 to the power 64) and time-to-time generation and assignment of a new interface ID. An other method of assigning the interface ID is by means of DHCP or manual configuration, as in IPv4.

Link-Local Address   A link-local address is a type of IPv6 unicast address that is configured automatically by the IPv6 stack. The link-local prefix is FE80::/10, followed by the interface ID in EUI-64 format. Link-local addresses are used for several basic IPv6 operations, such as the IPv6 neighbor discovery (ND) protocol. They are also used by IPv6 routing protocols to form neighbor relationships on the same links. Link-local addresses are commonly used to connect devices on the same link-local network without the need for global or unique-local addresses. Nodes on a local link can use link-local addresses to communicate with each other without the need for a router. Figure 2.4 shows the structure of a link-local address.

Unique-Local Address   The unique-local address (RFC4193) is a type of unicast address that is globally unique and is used for local communications within a site or within a limited set of sites. Unique-local addresses are not expected to be routable on the global Internet but to be routable inside a site or

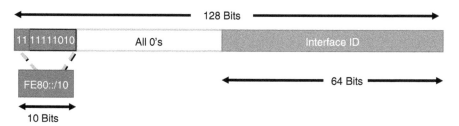

FIGURE 2.4   Structure of a link-local address.

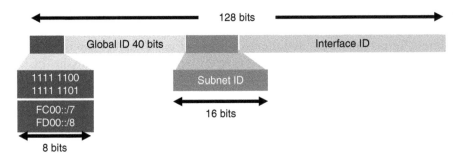

FIGURE 2.5   Structure of a unique-local address.

a limited set of sites. These addresses are unique in that they are a 40-bit global identifier to create a globally unique prefix. The global identifier is usually a pseudonumber. There are two ways to assign the unique-local prefix (FC00::/7, assigned centrally, and FD00::/8, assigned locally). The 16-bit subnet field could be used by individual organizations to create their own local addressing hierarchy and to identify up to 65,535 individual subnets followed by a 64-bit interface identifier. See Figure 2.5 for the structure of a unique-local address.

Loopback and Unspecified Addresses   In IPv4 the loopback address is 127.0.0.1. IPv6 also has a loopback address concept, 0:0:0:0:0:0:0:1. This loopback address could be compressed and represented as ::1. IPv6 also has an unspecified address, concept which is used as a placeholder when no address is available, such as during the duplicate address detection (DAD) process. The unspecified address is represented by 0:0:0:0:0:0:0:0, which could be compressed and represented as a double colon (::).

Mapped and Compatible Addresses   The IPv4 mapped and compatible addresses are formed by using the 32-bit IPv4 address and making it look like a 128-bit hexadecimal IPv6 address. These addresses are both used for specific purposes in IPv6 operations. For instance, IPv4 mapped addresses are used by the multiprotocol border gateway protocol (MP-BGP), and IPv4-compatible addresses are used during IPv6 testing, supporting automatic tunnels over IPv4 migration techniques. In the following IPv4 mapped address example, all zeros are followed by FFFF and finally a 32-bit IPv4 address, which is converted into hex format. Where C0 is 192, A8 is 168, and so on.

```
0:0:0:0:0:FFFF:192.168.30.1 = ::FFFF:C0A8:1E01
```

In the following IPv4 compatible address example, all zeros are followed by 32 bits of IPv4 address, which is converted into hex format. Where C0 is 192, A8 is 168, and so on.

```
0:0:0:0:0:0:192.168.30.1 = ::192.168.30.1 = ::C0A8:1E01
```

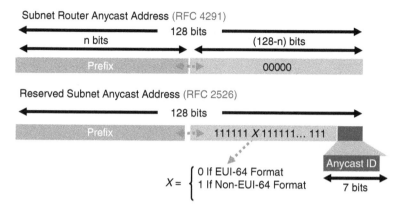

FIGURE 2.6   Structure of an anycast address.

2.1.3.2 Anycast Address   The concept of anycast is similar in IPv6 and IPv4. For IPv6, it is defined as a way to send a packet to the nearest interface that is a member of the anycast group. This enables a type of discovery mechanism to the nearest point: for example, to discover a SIP server. Currently, few anycast addresses are assigned. Examples are the subnet router anycast address and the reserved subnet router anycast address. The reserved subnet router anycast address is formed by the unicast address prefix at the high-order bits, followed by a series of ones and terminated by 7 bits on the lowest-order bits of the address. The last 7 bits are used to identify a specific anycast "group." The middle series of ones has an exception when the address type of the lower 64 bits is a EUI-64. Since the anycast address formed is not guaranteed to be unique, the EUI-64 uniqueness bit must be set to zero. Figure 2.6 shows the structure of an anycast address.

2.1.3.3 Multicast Address   An IPv6 multicast address (RFC3513) has a prefix FF00::/8 (1111 1111). The second octet defines the lifetime and scope of the multicast address, followed by the multicast group identifier. If the P bit is set, it indicates that the multicast address is based on the network prefix (RFC3306). The T bit is also called the lifetime bit. When the T bit is not set, it indicates a permanent "well-known" multicast address assigned by the global Internet numbering authority. Several well-known multicast addresses are used for IPv6 basic operations, the ND protocol, routing protocols, and other basic functionalities. When the T bit is set, it indicates a temporary multicast address. This is followed by four scope bits (Figure 2.7). Based on how the scope bits are set, they indicate how far this multicast packet could go. Some of the common scopes are 1 = node, 2 = link, 5 = site, 8 = organization, and E = global. Note that since IPv6 does not have a broadcast address, some of these scopes could be used to get a similar functionality. This is followed by a 112-bit group

FIGURE 2.7   Structure of a multicast address.

TABLE 2.1   Some Well-Known Multicast Addresses

Address	Scope	Meaning
FF01::1	Node-local	All nodes
FF02::1	Link-local	All nodes
FF01::2	Node-local	All routers
FF02::2	Link-local	All routers
FF05::2	Site-local	All routers
FF02::1:FFXX:XXXX	Link-local	Solicited-node

identifier. Some well-known multicast addresses are listed in Table 2.1. A complete list of well-known multicast addresses is given in RFC4291.

Another type of multicast address is the solicited-node multicast address, whose structure is shown in Figure 2.8. Whenever a unicast or an anycast address is configured on the interface of a device, a corresponding solicited-node multicast address is generated automatically and assigned to that interface by the IPv6 stack. This solicited-node multicast address is scoped to the local link and used for two fundamental IPv6 mechanisms. Since there is no concept of ARP in IPv6, the solicited-node multicast addresses is used by nodes and routers to learn the link layer address of the neighbor nodes and routers on the same local link. The other important function for which a solicited-node multicast address is used is in the DAD process. When a node plans to assign itself an IPv6 address, it sends a DAD message to make sure that no one else on the same link has assigned the same address. The solicited-node address is a well-known address and the first 104 bits are always FF02:0:0:0:0:1:FF, followed by the lower 24 bits of the corresponding link-local, unicast, or anycast address of the node.

Figure 2.9 shows the output of an Ethernet interface of an IPv6-enabled router. There are a few very important concepts to learn here. The link-local address (FE80::/10) and its corresponding solicited-node multicast address are assigned automatically by the IPv6 stack. Multicast all-nodes (FF02::1) and multicast all-routers (FF02::2) addresses have also been assigned automatically by the stack.

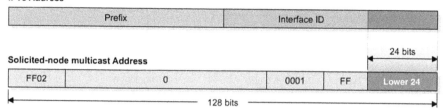

FIGURE 2.8 Structure of a solicited-node multicast address and its mapping.

```
R1#show ipv6 interface ethernet0
Ethernet0 is up, line protocol is up
  IPv6 is enabled, link-local address is FE80::200:CFF:FE3A:8B18
  No global unicast address is configured
  Joined group address(es):
    FF02::1
    FF02::2
    FF02::1:FF3A:8B18
  MTU is 1500 bytes
  ICMP error messages limited to one every 100 milliseconds
  ICMP redirects are enabled
  ND DAD is enabled, number of DAD attempts: 1
  ND reachable time is 30000 milliseconds
  ND advertised reachable time is 0 milliseconds
  ND advertised retransmit interval is 0 milliseconds
  ND router advertisements are sent every 200 seconds
  ND router advertisements live for 1800 seconds
  Hosts use stateless autoconfig for addresses
```

FIGURE 2.9 Command output for an IPv6-enabled interface on a Cisco router.

2.1.3.4   Reserved Address   The internet assigned numbers authority (IANA) has allocated certain IPv6 addresses and ranges for specific uses. For instance, 2000::/3 is reserved for IPv6 global unicast addresses, and FF00::/8 is reserved for IPv6 multicast addresses. For a complete list of currently allocated IPv6 addresses and ranges, refer to the IANA specifications.

2.1.4   ICMPv6

Internet control message protocol (ICMP) in IPv6 (RFC2463) functions the same as ICMP in IPv4 (RFC792) and has the same message types and codes, with minor differences. ICMPv6 (ICMP for IPv6) generates messages, such as ICMP destination unreachable, ICMP echo request, and ICMP echo reply messages. Similar to ICMPv4, ICMPv6 is often blocked by security policies implemented in corporate firewalls because of attacks based on ICMP. Special care needs to be taken while blocking ICMPv6 packets, as ICMPv6 is used by the essential protocols and operations of IPv6, such as the IPv6 neighbor discovery (ND) process, path MTU discovery (PMTUD), and the multicast

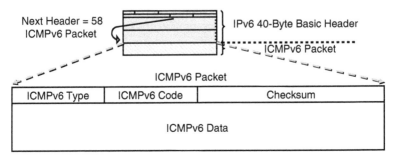

FIGURE 2.10   ICMPv6 header.

listener discovery (MLD) protocol. The MLD (RFC2701) protocol in IPv6 provides functionality similar to that of the IPv4 Internet group management protocol (IGMP).

An ICMPv6 packet follows either the IPv6 header or the EH (if used) and is the last piece of information in the IPv6 packet (Figure 2.10). A value of 58 in the Next Header field of the basic IPv6 packet header or EH identifies an ICMPv6 packet. The ICMPv6 Type and ICMPv6 Code fields identify ICMPv6 packet specifics, such as the ICMP message types. ICMPv6 has the following message types:

- Destination unreachable (type 1)
- Packet too big (type 2)
- Time exceeded (type 3)
- Parameter problem (type 4)
- Echo request (type 128)
- Echo reply (type 129)

2.1.5   Neighbor Discovery

The IPv6 neighbor discovery (ND) protocol (RFC4861, RFC4862) and ICMPv6 are at the heart of IPv6 operations. The ND protocol is built on top of the ICMPv6. All the ND messages are in essence ICMPv6 messages. IPv6 does not have ARP but rather, uses the ND process, which uses ICMPv6 messages and solicited-node multicast addresses to determine the link-layer address of a neighbor on the local link. ND is also used to verify the reachability of a neighbor and to keep track of neighbor routers, and supports autoconfiguration of addresses, duplicate address detection (DAD), and other IPv6 functionalities.

Following are five ND messages:

1. Router solicitation (RS) (ICMPv6 type 133)
2. Router advertisement (RA) (ICMPv6 type 134)

3. Neighbor solicitation (NS) (ICMPv6 type 135)
4. Neighbor advertisement (NA) (ICMPv6 type 136)
5. Redirect (ICMPv6 type 137)

RS messages are sent by booting nodes to request RA messages for configuring the interfaces as part of the stateless address autoconfiguration (SLAAC) (RFC4862) process. Routers also send periodic RA messages to the all-nodes multicast address. All the messages are exchanged using link-local and well-known multicast addresses.

Let's take a look at the RS and RA message exchange shown in Figure 2.11. The host uses its link-local address to source the RS, with FF02::2 as the destination address for this message. The router responds with a RA messages using its link-local address as the source address and the FF02::1 as the destination address. The RA message contains three types of flags: A, the autonomous address-configuration flag; M, the managed address configuration flag, and O, the other configuration flag, which determine the methodology for obtaining an IPv6 address and other network parameters by the host. If the RA message contains the bits A = 1, M = 0 and O = 1, the host uses SLAAC for IPv6 address assignment and stateless DHCPv6 for obtaining other network parameters. If the RA message contains bits A = 0, M = 1, and O = 1, the host uses stateful DHCPv6 for address assignment as well as for obtaining other network parameters.

Before the host can configure itself with an IPv6 address, it needs to make sure that the address is unique and is not being used by another host or device on that link. To verify that the IPv6 address is unique, the host sends out a duplicate address detection (DAD) message on the link. The DAD request is essentially a neighbor solicitation (NS) message with a source address of (::) or unspecified, and a destination address equal to the solicited-node multicast address for the IPv6 address. If the host does not get a reply back to this NS message, it will assume that the address is unique and assign the address to itself. If the host gets a neighbor advertisement (NA) message in response to the

IPv6 Host	Dual-Stack Router
1. RS	2. RA

1. ICMPv6 Type = 133 (RS)
Src = Link-local address
(FE80::207:50FF:FE5E:9460)
Dst = All-routers multicast address (FF02::2)
Query = Please send RA

2. ICMPv6 Type = 134 (RA)
Src = Link-local address
(FE80::209:50FF:FE5E:6450)
Dst = All-nodes multicast address (FF02::1)
Data = Options, IPv6 prefix, lifetimes, Auto-configuration flag

FIGURE 2.11   RS and RA message exchange.

FIGURE 2.12   Access BB network topology.

NS, it knows that the address is already in use and it cannot assign the address to itself. According to RFC4862, the host must not assign this address to its interface and should log a system management error message. If a duplicate address is detected for a link-local address that is being derived by using EUI-64 format, IPv6 operation should be disabled on this interface. The network administrator needs to resolve the IPv6 address conflict and enable this interface manually. The host performs a DAD check for each address (link-local, global unicast, unique-local) that it tries to assigns itself. DAD is also used by routers before they assign an IPv6 address to their interfaces. See Figure 2.16 for DAD operation.

Figure 2.12 analyzes a typical topology of access broadband networks, with an NSP provisioning system, NSP edge router (ER), customer premises gateway router (GWR), and a host connected to the GWR. In this scenario the GWR is playing a dual role, acting as a DHCPv6 client to the ER and as a DHCPv6 server (with minimal functionality) for the host connected to it.

Table 2.2 is based on the topology in Figure 2.12 and shows the source and user of the RA, indicates different options for setting the three bits (A, M, and O) in the RA messages, and resulting operations. Depending on how these bits are set, the host may assign itself an IPv6 address using the SLAAC methodology.

NS messages are sent on the local link when a node wants to determine the link-layer address of another node on the same local link. This functionality is similar to the ARP in IPv4, but unlike IPv4 ARP, IPv6 uses well-known multicast addresses instead of broadcast addresses. The IPv6 NA message is a

TABLE 2.2   Operation Based on A/M/O Bits in RA

Source of RA	User of RA	A	Operation	M/O	Operation
			A-Bit		M/O Bits
ER	GWR E1	0	Don't do stateless address assignment	1/1	Use DHCPv6 for address + other configuration (i.e., stateful DHCPv6)
GWR	Host	1	Do stateless address assignment	0/1	Use DHCPv6 for other configuration (i.e., stateless DHCPv6)

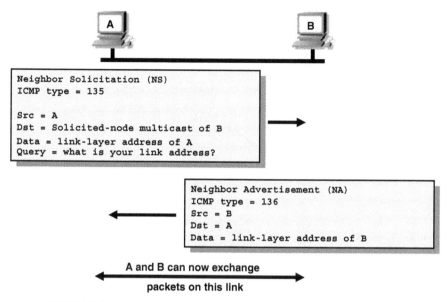

FIGURE 2.13 NS/NA message exchange during address resolution.

response to the NS message (Figure 2.13). Once the NS message is received, the receiving node replies by sending a NA message on the local link. After the exchange of NS and NA, the source and destination nodes can begin to communicate.

To elaborate further, node A sends NS to discover the Layer2 (L2) address of node B, based on the target's IPv6 address. Figures 2.14 and 2.15 show a successful NS and NA message exchange and subsequent discovery of the L2

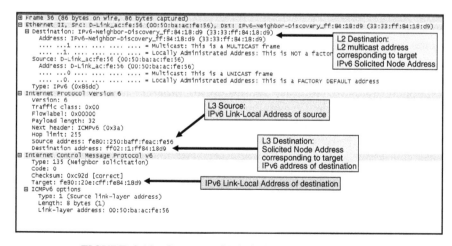

FIGURE 2.14 Contents of NS during address resolution.

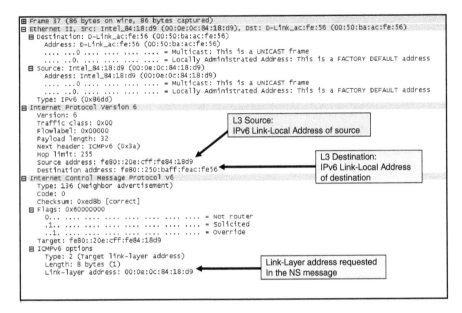

```
⊞ Frame 37 (86 bytes on wire, 86 bytes captured)
⊟ Ethernet II, Src: Intel_84:18:d9 (00:0e:0c:84:18:d9), Dst: D-Link_ac:fe:56 (00:50:ba:ac:fe:56)
  ⊟ Destination: D-Link_ac:fe:56 (00:50:ba:ac:fe:56)
      Address: D-Link_ac:fe:56 (00:50:ba:ac:fe:56)
      .... ...0 .... .... .... .... = Multicast: This is a UNICAST frame
      .... ..0. .... .... .... .... = Locally Administrated Address: This is a FACTORY DEFAULT address
  ⊟ Source: Intel_84:18:d9 (00:0e:0c:84:18:d9)
      Address: Intel_84:18:d9 (00:0e:0c:84:18:d9)
      .... ...0 .... .... .... .... = Multicast: This is a UNICAST frame
      .... ..0. .... .... .... .... = Locally Administrated Address: This is a FACTORY DEFAULT address
      Type: IPv6 (0x86dd)
⊟ Internet Protocol Version 6
      Version: 6
      Traffic class: 0x00
      Flowlabel: 0x00000
      Payload length: 32
      Next header: ICMPv6 (0x3a)
      Hop limit: 255
      Source address: fe80::20e:cff:fe84:18d9
      Destination address: fe80::250:baff:feac:fe56
⊟ Internet Control Message Protocol v6
      Type: 136 (Neighbor advertisement)
      Code: 0
      Checksum: 0xed8b [correct]
  ⊟ Flags: 0x60000000
      0... .... .... .... .... .... .... .... = Not router
      .1.. .... .... .... .... .... .... .... = Solicited
      ..1. .... .... .... .... .... .... .... = Override
      Target: fe80::20e:cff:fe84:18d9
  ⊟ ICMPv6 options
      Type: 2 (Target link-layer address)
      Length: 8 bytes (1)
      Link-layer address: 00:0e:0c:84:18:d9
```

L3 Source:
IPv6 Link-Local Address of source

L3 Destination:
IPv6 Link-Local Address
of destination

Link-Layer address requested
in the NS message

FIGURE 2.15   Contents of NA during address resolution.

address of node B. Note the main contents of the NS and NA messages and the L2 and Layer3 (L3) source and destination addresses shown in the sniffer traces in the Figures.

To further analyze the DAD process, let us observe the main contents of the NS and NA messages and the L2 and L3 source and destination unicast and multicast addresses in Figure 2.16. The NS and NA messages are divided into three portions: Ethernet header, IPv6 header, and NS or NA header. In the example shown in Figure 2.16, host A uses DAD to verify the existence of a duplicate address before assigning the address FE80::2:260:8FF:FE52:F9D8 to its interface by sending a NS message on the local link. The first piece of information is listed under the Ethernet header, where the destination IPv6 Ethernet MAC address is listed. The IPv6 Ethernet MAC address is built by mapping an IPv6 multicast address to an Ethernet address as follows: 33:33:<last 32 bits of the IPv6 multicast address>. In this case it is 33-33-FF-52-F9-D8. The second piece of information is listed under the IPv6 header, where the IPv6 L3 source address is the unspecified address (::), since host A has not assigned yet the IPv6 address. The destination address is the solicited node multicast address (FF02::1:FF52:F9D8), built from the target link-local address (FE80::2:260:8FF:FE52:F9D8). The third piece of information in listed under the NS header, with the host B L3 link-local address (FE80::2:260:8FF:FE52:F9D8) as the target address.

In response to the NS message, an NA message is sent. Let us also observe the contents of NA. The first piece of information, 33-33-00-00-00-01

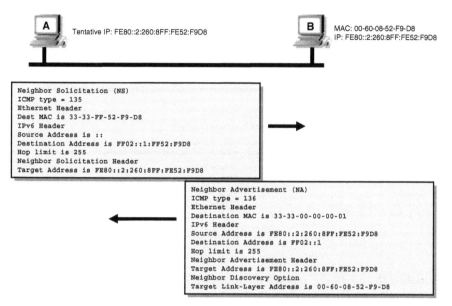

FIGURE 2.16   NS/NA message exchange during DAD.

(the all-node multicast address at L2), is listed under the Ethernet header. The second piece of information is listed under the IPv6 header, where the IPv6 L3 source address is host B's link-local address, FE80::2:260:8FF:FE52:F9D8. The IPv6 L3 destination address is FF02::1 (the all-node multicast address at L3). The third piece of information in listed under the NA header; where the target address is the host B L3 link-local address, FE80::2:260:8FF: FE52:F9D8, as this is the address for which the NS was sent initially. Here also notice that an NA option has been appended to the NA message where host B's 48-bit MAC address has been added. Not a common scenario, but in this example host B has a duplicate address that host A was planning to assign. According to the standard, at this point host A should disable its interface and the network administrator should further investigate the cause of the duplicity. If the host does not receive any NA in response to the NS, it will assume that there is no duplicity and will go ahead and assign itself an address.

As with IPv4, an IPv6 redirect message is sent by a router only to help with the reroute of a packet to a better router. The node receiving the redirect message will then readdress the packet to a router that has a better path to the destination. Routers send redirect messages only for unicast traffic, and only to the originating nodes. In the example shown in Figure 2.17, node A sends a normal IPv6 packet to its default router R2 with a destination of 2001: db8:C18:2::1. R2 cannot forward this packet but knows that R1 has an optimal route to this destination. R2 sends a redirect message back to node A,

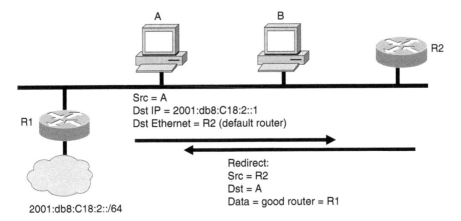

FIGURE 2.17   Use of RA for the redirect process.

suggesting the use of R1 to reach the targeted destination. Node A then resends the packet using R1 as its default router for the destination address.

Let's consider a scenario where customer A has received its IPv6 address range from service provider X (SP-X), and customer A decides to change its service provider to SP-Y. In this case, customer A will receive a new IPv6 address range from SP-Y. Once a new IPv6 address range is received from SP-Y, customer A will have to renumber its network with the new address range received from SP-Y. Renumbering is made simpler in IPv6 by configuring the routers to send out an RA with the added prefix information (Figure 2.18). This information includes a current prefix with a shorter lifetime, so it is not used after the lifetime expires, and a new prefix with a normal lifetime. When the old prefix is no longer usable, the RA will only include the new prefix (RFC4861). Again, link-local and well-known multicast addresses are used in the process.

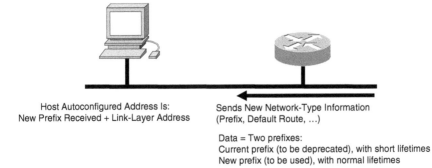

FIGURE 2.18   Use of RA during a renumbering process.

2.1.6 Secure Neighbor Discovery

In Section 2.1.5, we discussed the ND protocol (RFC4861, RFC4862) and some of its main functions and uses. Some uses of the ND protocol are to discover the link-layer addresses of other devices, to discover routers, and to maintain reachability information about paths to active neighbors. As in IPv4 ARP, some of the uses of ND raise issues of vulnerability, threats, and obvious security concerns. To address these concerns, IETF defined the secure neighbor discovery protocol (SEND) in RFC3971, together with cryptographically generated addresses (CGAs) in RFC3972. RFC3972 specifies a method for securely associating a cryptographic public key with an IPv6 address in the SEND protocol. To learn about ND vulnerabilities, threats, and security concerns, readers are referred to RFC3756, "IPv6 Neighbor Discovery Trust Models and Threats".

Although the SEND protocol addresses some security concerns, it also has some caveats. For instance, a private/public key pair is required on all devices for CGAs, which adds processing overhead on routers, as they end up doing excessive public/private key calculations and maintaining more states. It has been suggested that some of the public/private key calculations could be done in advance so that routers do not have to perform these calculations on the fly, but that still exposes these devices to potential DoS targets. At the time of this writing, SEND is available only on the Linux operating system.

2.1.7 Provisioning

IPv6 was never intended to be revolutionary in its design. Its features were driven mainly by the shortcomings of IPv4, particularly the latter's limited address space. IPv6 was focused primarily on developing a larger address space and providing reasonable improvements over IPv4 based on different operational requirements, such as huge address range, long-lived connections, and permanent prefixes. As large IPv6 broadband access deployments become common worldwide, the industry, SPs, and standards bodies have stepped up to define extensions to the protocol and introduced changes to support provisioning protocols and methodologies for scaling and facilitating large IPv6 broadband access deployments. These extensions encompass newer IPv6 address assignment methodologies: domain name server (DNS); trivial file transfer protocol (TFTP); authentication, authorization, and accounting (AAA); RADIUS extensions; definition of dynamic host configuration protocol version 6 (DHCPv6); and others. In this section we cover changes and extensions to IPv6-related protocols and discuss issues relating to IPv6 operations and access design choices. Table 2.3 compares the key provisioning features of IPv4 and IPv6, such as address assignment, address resolution, router discovery, and name resolution.

2.1.7.1 IP Address Allocation As you may recall, in the IPv4 provisioning environment, one method of address assignment is to allocate address pools by

TABLE 2.3   IPv4/IPv6 Provisioning Comparison

Function	IPv4	IPv6
Address assignment	DHCPv4	DHCPv6, SLAAC, reconfiguration
Address resolution	ARP, RARP	NS, NA, inverse ND RFC3122
Router discovery	ICMP router discovery	RS, RA
Name resolution	DNS for IPv4	DNS for IPv6

broadband providers. When a client comes online, an IPv4 address is assigned to it from a predefined address pool. When the client goes offline, IPv4 address is released and may be assigned a different client. Another IPv4 provisioning practice is to use network address translation (NAT) extensively within the remote site. Besides the several other benefits of IPv6, one key objective of the protocol IPv6 was to provide more IP addresses to the Internet community. Since a huge number of IPv6 addresses are available, IPv6 can assign long-lived addresses, hence eliminating the need for shared address pools by means of DHCP and NAT. In IPv4 provisioning practices, the end user could be assigned with one address, whereas in IPv6 a single link is assigned with a /64 prefix (the smallest address block allocation). The IAB/IESG recommendations on IPv6 address allocation to sites listed in RFC3177 states: "... It recommends the assignment of /48 in the general case, /64 when it is known that one and only one subnet is needed." At the time of writing this recommendation is being reconsidered in standards bodies and the networking industry in general. The current typical IPv6 allocation is a /48 prefix followed by a 16-bit SLA field which allows the end site to have more than 65,000 subnets in its network. We can assign multiple IPv6 addresses to a host. A common example is a host with a global unicast address, a unique-local address, and a link-local address.

The escalated need for assigning more permanent static addresses is due to an exponential increase in the number of servers and other endpoints worldwide. Here the use of temporary addresses is not practical, since these servers and endpoints need to be accessed via DNS entry, and if the address is temporary, these DNS entries need to be changed whenever a new temporary address is assigned. In IPv4 this was achieved by configuring DNS and the port-mapping features of NAT, which resulted in other operational complexities.

IPv4 does not offer any methodology for assigning multiple addresses automatically to remote sites. There are two dominant methodologies for address assignment in IPv4: point-to-point protocol (PPP) and DHCPv4. Due to the lack of IPv4 addresses, the Internet protocol control protocol (IPCP) was not required to carry and assign multiple addresses, and DHCPv4 was never built or enhanced to provide multiple addresses to remote sites.

IPv6 introduces the concept of SLAAC and DHCPv6 prefix delegation (PD). In IPv4, clients typically receive a single IP address from the DHCPv4 server or the address is configured statically. In IPv6 a single client can have multiple addresses assigned to it. An IPv6-enabled GWR may receive a /48 to /64 prefix from the SP, depending on the SP's policy and the customer's requirements. If the GWR has multiple networks connected to its interfaces, it can receive a /48 prefix from the provider and uses the same /48 to further carve out /64 prefixes to each of its interfaces. Hosts connected to these interfaces can configure themselves automatically with an IPv6 address using the /64 prefix. As discussed previously, there are four ways to append an interface ID (the last 64 bits of the IPv6 address) to make it a complete 128-bit IPv6 address: namely, the EUI-64 format, manually configured and randomly generated (RFC3041), and DHCPv6.

While the initial IPv6 deployments relied primarily on PPP for IPv6 address allocations, newer and larger-scale IPv6 deployments are getting away from PPP-based access models as the benefits of IPv6 ND, DHCPv6, and its associated methodologies for provisioning of IPv6 addresses and other network parameters are now better understood. IPv6 uses primarily two ways to assign addresses, automatically, discussed next.

Stateless Address Autoconfiguration   Along with DHCP; PPP is one of the most commonly used techniques in IPv4 to connect and assign automatic addresses to Internet users. The initial IPv6 deployments took the same route while using PPP to achieve the same functionality. However, the functionality of IPv6CP differs from IPCP (IPv4CP), as it negotiates only a unique identifier, while ND is used for address assignment. Once the PPP link is established and a unique identifier is negotiated, the address prefix is sent in an RA message. The address prefix sent in the RA message could be obtained through manual configuration from an address pool or AAA server.

SLAAC (RFC4862) enables basic configuration of the IPv6 interfaces in the absence of a DHCPv6 server. SLAAC relies on the information in the RA messages to configure the interface. Within the RA message there are certain bits that dictate the actions taken by the receiver of an RA message. As shown in Table 2.2, the RA message contains a bit called the A-bit. When the A-bit is set, the receiving device engages the SLAAC mechanism. The /64 prefix included in the RA is used as the prefix for the interface address. For Ethernet, the remaining 64 bits are obtained from the interface ID in the EUI-64 format. Thus, an IPv6 node can autoconfigure itself with a globally unique IPv6 address by appending its link-layer address based on the interface ID built in the EUI-64 format to the prefix provided in the RA. Figure 2.19 illustrates the SLAAC process. Randomly generating interface ID as described in RFC3041 is part of stateless autoconfiguration and is used to address some security concerns.

DHCPv6   The operational model of DHCPv6 is based on DHCPv4 with some changes and new concepts to grasp. Lots of work continues in standards bodies to enrich the functions of DHCPv6 (RFC3315) and toward providing scalable

SUBNET PREFIX Received
+ Interface ID built from EUI-64

ICMP Type = 134 (RA)
SUBNET PREFIX Advertised

FIGURE 2.19   SLAAC process.

deployment models. Some highlights of DHCPv6 and its features are listed below.

- As we know, link-local address, are used by different protocols and functionalities; in DHCPv6, clients use the link addresses for message exchanges.
- In DHCPv4, clients use broadcast addresses to reach the DHCPv4 server. Since IPv6 does not support the concept of broadcast address, multicast addresses are used to achieve the same functionalities. The following two important multicast addresses are used commonly:
  - Clients use the link-scoped multicast address All_DHCP_Relay_ Agents_and_Servers (FF02::1:2) to communicate with neighboring relay agents and servers which are members of this multicast group.
  - Relay agents use a site-scoped multicast address All_DHCP_Servers (FF05::1:3) when sending messages to all servers or when the unicast addresses of the servers are not known. All servers within a site are members of this multicast group.
- Clients listen for DHCP messages on UDP port 546. Servers and relay agents listen for DHCP messages on UDP port 547.
- DHCPv6 message names have been changed, although the functionality stays the same, as shown in Table 2.4.
- DHCPv6 devices such as clients and servers are identified by the DHCP unique identifier (DUID). The DUID is carried as a DHCPv6 option and is unique.

TABLE 2.4   Comparison of Basic DHCPv4 and DHCPv6 Message Types

DHCP Messages	IPv4	IPv6
Initial message exchange	4-way handshake	4-way handshake
Message types	Broadcast, unicast	Multicast, unicast
Client → server (1)	DISCOVER	SOLICIT
Server → client (2)	OFFER	ADVERTISE
Client → server (3)	REQUEST	REQUEST
Server → client (4)	ACK	REPLY

- The DHCPv6 server can assign multiple IPv6 prefixes to each client by identity associations.
- DHCPv6 supports a stateful configuration (RFC3315), where the DHCPv6 server assigns IPv6 addresses and provides other network parameters to clients. However, currently, very few IPv6 hosts implementations support this as stateless autoconfiguration is preferred.
- DHCPv6 also supports stateless configuration (also called DHCPv6lite) (RFC3736). The DHCPv6 server does not assign addresses but, instead, provides configuration parameters, such as DNS server information, to clients while an address assignment is carried out using SLAAC.
- IPv6 introduces the new concept of DHCPv6 prefix delegation (PD) (RFC3633), which allows routers to be delegated to assign prefix(es) to other requesting routers or other devices on customer premises. Requesting routers use DHCPv6 options to request prefix(es) from the delegating router.

DHCPv6 Prefix Delegation   The DHCPv6 PD (RFC3633) provides a scalable solution to SPs for managing large numbers of users (remote sites), where each user may have multiple links or devices in its own network. In this provisioning approach the NSP assigns and keeps track of a permanent and shorter than /64 prefix to the remote sites. Figure 2.20 illustrates a common access layer of NSP network and its components and proposed IPv6 provisioning technique. In this particular scenario, a series of events take place in provisioning the remote site and eventually assigning IPv6 addresses to hosts connected to the remote sites. The ER–GWR link needs to get global IPv6 addresses assigned; this could be achieved via enabling PPP on the ER–GWR link. PPP is used here, but other methods could be used as well. While establishing a PPP link, during the link control protocol (LCP) negotiation, the GWR's username and password need to be authenticated. The ER forwards the authentication request to the provisioning servers located in the NSP network. The provisioning server is preconfigured for this username/password to authenticate. Once authentication has been completed, the provisioning server will return a /64 prefix to the ER. As explained earlier, IPv6CP does not carry the prefix information. This /64

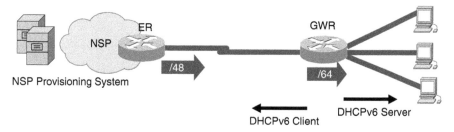

FIGURE 2.20   NSP edge network and ipv6 provisioning strategy.

prefix will be sent to the GWR on the ER–GWR link using the ND RA message. The /64 prefix is also injected in the NSP routing system. After the GWR assigns itself an IPv6 address, it will send the DHCPv6 SOLICIT message to the ER using the option request option (ORO). In this ORO, the GWR will request PD (option 25) and DNS (option 6) information from the ER.

A complete list of DHCPv6 options is listed in the appendix to this book. The ER is acting as a DHCPv6 relay, and as soon as it receives the SOLICIT message it forwards the SOLICIT message (RELAY-FORWARD with SOLICIT) to the DHCPv6 requesting router (i.e., GWR). Figure 2.21 illustrates DHCPv6 message exchange when a DHCPv6 relay is present in the topology for relaying the messages between client and server. The DHCPv6 server returns the information requested by sending a RELAY-REPLY message to ER with an ADVERTISE message. This message would have PD (/48 prefix), domain server, and domain list information requested in the SOLICIT message. Since the DHCPv6 server is preconfigured with a permanent address for this particular GWR, the SP will be maintaining the information about which prefix is assigned to this GWR. This predefined prefix will be injected in the SP routing scheme to make sure that no routing anomalies exist and that SP knows exactly where this particular GWR is located.

At this point the ER will send an ADVERTISE message with PD (/48 prefix), domain server, and domain list information to the GWR. The GWR will send a REQUEST message back to the ER with the same information that it received in the ADVERTISE message, for verification purposes. Since the ER is acting as a DHCPv6 relay, it will forward this REQUEST message (RELAY-FORWARD with REQUEST) to the DHCPv6 server. The DHCPv6 server will respond back to the ER with a REPLY (RELAY-REPLY with a REPLY) message. The ER will send this REPLY message to the

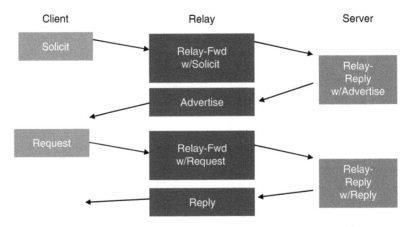

FIGURE 2.21 DHCPv6 messages when the DHCPv6 relay agent is present.

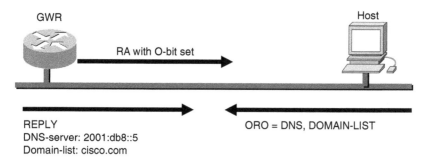

FIGURE 2.22   GWR–host provisioning message exchange.

GWR. The REPLY message will contain the same information that was present in the ADVERTISE message (/48 prefix, domain name, domain list). The complete ER-GWR DHCPv6 message exchange for this scenario is presented in Chapter 6.

Once the GWR receives the REPLY message with the /48 prefix, the GWR will carve the /64 prefixes out of the /48 prefix. This functionality requires specific configuration on the GWR, which is presented in Chapter 5. These /64 prefixes will be assigned to the links connected to the GWR. At this point any host connected to the GWR will use the ND/SLAAC process to acquire the /64 prefix and assign itself an address (global or unique local). Once the address is assigned by the host and the O-bit is set (part of the RA message), it sends a DHCv6 INFORMATION-REQUEST message, with ORO set to DNS. This request will be accepted by the GWR (which has already cashed the domain name and domain list information in the REPLY message sent from the ER) in response to the information requested.

As shown in Figure 2.22 the GWR is acting as a DHCPv6 client of its upstream neighbor (i.e., the ER) and is acting as a basic DHCPv6 server for its downstream neighbors (i.e., host connected to itself). This final process of address assignment and acquisition of the domain name and domain list information by the host makes this a unique and scalable end-to-end provisioning solution specific to IPv6. Here a stateful DHCPv6 option could be used when no provisioning server is used, but rather, all the prefix information is maintained on the ER by using the DUID to keep track of the prefixes assigned to each GWR, a less preferred and unscalable option.

2.1.7.2 DNS   Renumbering requires changes in the DNS entries and the introduction of new IPv6 DNS records. Renumbering an entire site also requires that all the routers be renumbered. Stateless autoconfiguration does not address the issue of finding the DNS server for DNS resolution (stateless DHCP does) or registering the computer in the DNS space—dynamic DNS is an option.

IPv6 introduces new DNS record types for IPv6 addresses that are supported in the DNS name-to-address and address-to-name lookup processes.

- An AAAA record, also known as a *quad A record*, is equivalent to an A record in IPv4. It maps a host name to an IPv6 address.
- A PTR record is equivalent to a pointer (PTR) record, which specifies address-to-host name mappings in IPv4. Inverse mapping used in IP address-to-host name lookup uses the PTR record. The top-level domain for IPv6 addresses is ip6.arpa.
- A6 and DNAME and binary labels records make renumbering easy for inverse mapping (IP address to host name).

2.1.7.3   AAA   New attributes have been added to support IPv6 functionality. AAA together with RADIUS is typically used in broadband access networks to provide authentication and authorization to users connecting to the SP's network. Current RADIUS attributes show IPv4 dependencies. They are encoded to support 32-bit addresses and run over an IPv4 transport. Although the IPv4 dial-up model assigns only IPv4 host addresses, IPv6 RADIUS attributes carry IPv6 addresses and make room for concepts such as IPv6 address prefix assignments. To allow the RADIUS protocol to run over an IPv6 transport, it is also desirable to support future IPv6-only deployments.

Several implementations of IPv6 on RADIUS support both vendor-specific attributes (VSAs) and RFC3162 attributes. The ISP RADIUS server requires an upgrade of the software that supports such RFC3162 attributes as framed-interface-Id, framed-IPv6-prefix, login-IPv6-host, framed-IPv6-route, and framed-IPv6-pool.

2.1.7.4   TFTP   TFTP is a standard protocol that uses UDP packets to transfer data from one point to another. TFTP is capable of supporting IPv4 and IPv6 through certain options. Several vendors have ported this functionality in their respective products to support IPv6.

2.1.7.5   NTPv4   Necessary changes have been made to support IPv6 in the NTPv4 specification, along with support for the IPv6 address family in addition to the IPv4 address family. Several vendors have ported the NTPv4 functionality into their respective products to support IPv6. Note that NTPv4 is the fourth version of NTP (network time protocol) and is not limited to IPv4.

2.2   SUMMARY

In this chapter we highlighted some of the basics of the IPv6 protocol, operations, addressing, and provisioning. We demonstrated that although IPv6 features some new concepts, similarities to IPv4 remain. Some of the new IPv6 concepts include a fixed 40-byte header, a 128-bit address space, no broadcast addresses, ND protocol, SLAAC, and DHCPv6 PD. This chapter focused on concepts of special interest to network engineers and architects involved in designing and deploying IPv6 in access broadband networks. We also provided the reader with several references for further exploration and to obtain specific details about the IPv6 and related protocols.

REFERENCES

1. S. Asadullah and A. Ahmed, "IPv6 in Broadband," Cisco Systems, Inc. Packet Magazine, Fourth Quarter 2004.
2. S. Asadullah, A. Ahmed, C. Popoviciu, P. Savola, and J. Palet, "ISP IPv6 Deployment Scenarios in Broadband Access Networks," RFC4779, January 2007.
3. R. Hinden, M. O'Dell, and S. Deering, "An Aggregatable Global Unicast Address Format," RFC2374, July 1998.
4. R. Hinden and S. Deering, "Internet Protocol, Version 6 (IPv6) Specification," RFC2460, December 1998.
5. IEEE, "Guidelines for 64-Bit Global Identifier (EUI-64) Registration Authority," http://standards.ieee.org/db/oui/tutorials/EUI64.html, March 1997.
6. T. Narten and R. Draves, "Privacy Extensions for Stateless Address Autoconfiguration in IPv6," RFC3041, January 2001.
7. S. Thomson, T. Narten, and T. Jinmei, "IPv6 Stateless Address Autoconfiguration," RFC4862, September 2007.
8. S. McFarland, "BRKIPM-2005: Enterprise IPv6 Deployment," Cisco Networkers 2008, January 2008.
9. A. Ahmed and S. Asadullah, "LABIPM-2002: IPv6 Labtorial," Cisco Networkers 2007, January 2007.
10. B. Lourdelet, "Application Note: IPv6 Access Services," Cisco Systems, Inc.
11. B. Lourdelet, "Application Note: DHCPv6," Cisco Systems, Inc.
12. Cisco Systems, Inc., tutorial on IPv6 basics: "The ABC of IPv6."
13. T. Narten, E. Nordmark, and W. Simpson, "Neighbor Discovery for IP Version 6," RFC2461, December 1998.
14. B. Aboba, G. Zorn, and D. Mitton, "RADIUS and IPv6," RFC3162, August 2001.
15. R. Droms et al. "Dynamic Host Configuration Protocol for IPv6 (DHCPv6)," RFC3315, July 2003.
16. O. Troan and R. Droms, "IPv6 Prefix Options for Dynamic Host Configuration Protocol (DHCP) Version 6," RFC3633, December 2003.
17. R. Droms, "DNS Configuration Options for Dynamic Host Configuration Protocol for IPv6 (DHCPv6)," RFC3646, December 2003.

3 Deploying IPv6 in Cable Networks

In this chapter we discuss IPv6 deployment in cable networks, providing an overview of key network elements, different deployment models, and challenges in deploying IPv6 over a cable network.

Most equipment used in cable networks today conforms to the data over cable service interface specifications (DOCSIS) standard. DOCSIS defines the communications and operations support interface requirements for data over cable systems. The specification permits high-speed data transmission over an existing cable TV (CATV) network. It is employed by many multiple system operators (MSOs) to provide Internet access over their existing hybrid fiber coaxial (HFC) infrastructure. The first DOCSIS specification, version 1.0, which was published in March 1997, provided best effort (BE) data services to end users. DOCSIS 1.1, which was released in April 1999, allowed MSOs to deploy real-time services, such as IP telephony, by adding quality of service (QoS) capabilities to the upstream. In December 2001 the specification was again revised and released as DOCSIS 2.0, providing enhanced upstream transmission speeds. This catered for increased demand for symmetric services by business customers and a rise in peer-to-peer applications. DOCSIS 3.0, published in August 2006, specified increased transmissions speeds, this time for both upstream and downstream. This version also introduced support for several new features, including IPv6.

Products that comply with the DOCSIS standard are marked as DOCSIS certified or qualified. This ensures that cable operators can deploy products from different vendors in their network and that these products will interoperate. In the next section we discuss key network elements in a cable network.

3.1 CABLE NETWORK ELEMENTS

Here are some of the key elements of a cable network:

- *Hybrid fiber coaxial (HFC) plant* The HFC plant, also referred to as the cable plant, is the underlying transport for carrying subscriber traffic between the cable modem termination system, the cable modem, and hosts.

Deploying IPv6 in Broadband Access Networks, By Adeel Ahmed and Salman Asadullah
Copyright © 2009 John Wiley & Sons, Inc.

- *Cable modem termination system (CMTS)* The CMTS, located at the head end or distribution hub, provides data connectivity between the host/CM and other devices on the IP network.
- *Cable modem (CM)* This device is a modulator–demodulator at subscriber locations intended for transporting data traffic over the cable plant.
- *Multimedia terminal adapter (MTA)* The MTA transports VoIP traffic to and from subscribers.
- *Residential gateway router (GWR)* The GWR provides Layer 3 services to hosts.
- *Customer premise equipment (CPE)/Host* This is a PC or notebook appliance that is connected to the CM or GWR.
- *Edge router (ER)* The edge router, upstream of the CMTS, connects to the ISP backbone network. An ER can aggregate multiple CMTSs. The ER functionality can also exist on the CMTS.
- *Management servers* These servers are used to manage devices in a cable network, including CMTS, CM, MTA, and set-top boxes. Included are servers that poll various devices to retrieve information via SNMP and SYSLOG servers that collect logs from various devices.
- *Provisioning servers* Dynamic host configuration protocol (DHCP), domain name system (DNS), trivial file transfer protocol (TFTP), and time-of-day (ToD) servers are used for provisioning cable modems, MTA, and STB.

Figure 3.1 illustrates the key elements of a cable (DOCSIS) network. The figure shows three different deployment models:

1. The first model displays a CM/MTA connected to the CMTS over the HFC plant. The CM acts as a Layer 2 bridge and bridges data between the CPE devices and the CMTS. It has an embedded MTA for transporting VoIP traffic.
2. The second model displays a GWR connected to the CM/MTA. The GWR is used to provide Layer 3 services to hosts connected to it.
3. The third model displays a CM router with an embedded MTA and GWR functionality. This CM acts as a Layer 3 device and provides data as well as VoIP services to hosts connected to it.

These deployment models are discussed in more detail in later sections.

3.2 CABLE NETWORKS TODAY

The cable networks today are divided primarily into two categories: (1) bridged CMTS networks and (2) routed CMTS networks. Next we describe these environments in detail.

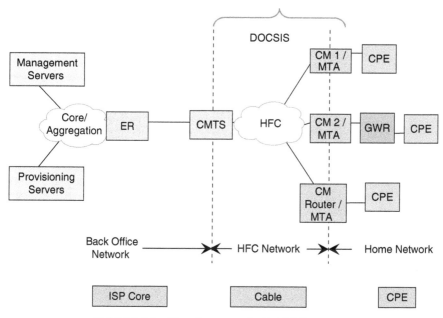

FIGURE 3.1   Key elements of a DOCSIS network.

3.2.1 Bridged CMTS Networks

In a bridged CMTS network, the CMTS and CM are Layer 2 devices and bridge all data traffic between the CPE devices and the ER. The ER then routes traffic through the ISP network to the Internet. The CM and CMTS support a certain degree of Layer 3 functionality for management purposes.

3.2.1.1 Deploying IPv6 in a Bridged CMTS Network   When deploying IPv6 in this environment, the CM and CMTS continue to operate as bridging devices, thereby providing a smooth transition and reducing operational complexity. The CM and CMTS bridge IPv6 unicast and multicast packets between the ER and the hosts. If there is a GWR connected to the CM, the GWR acts as a transit for IPv6 unicast and multicast traffic to and from the ER. IPv6 can be deployed in a bridged CMTS network either natively or via tunneling.

Figure 3.2 illustrates a bridged CMTS environment. In the first deployment scenario, the CM and CMTS remain Layer 2 devices and bridge both IPv4 and IPv6 traffic between the host and the ER. The host and ER are upgraded to support dual-stack functionality. The host uses the ER as its Layer 3 next hop.

In the second deployment scenario the CM and CMTS act as Layer 2 devices and bridge both IPv4 and IPv6 traffic between the GWR and the ER. The host, GWR, and ER will need to be upgraded to a dual-stack functionality. The host

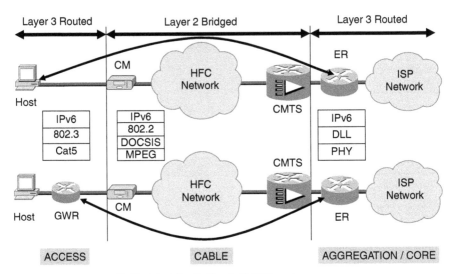

FIGURE 3.2   Bridged CMTS environment.

will use the GWR as its default gateway, and the GWR will use the ER as its Layer 3 next hop.

The CM and CMTS must be able to bridge native IPv6 unicast and multicast traffic. The CMTS must provide IP connectivity between hosts attached to CMs and must do so in a manner consistent with Ethernet-attached customer equipment. To do so, the CM and CMTS must forward neighbor discovery (ND) packets between the ER and the host or GWR attached to the CM. The CM and CMTS can distinguish IPv6 traffic from IPv4 traffic using the frame protocol ID (IPv6 = 0x86DD, IPv4 = 0x0800). This is necessary to provide the appropriate quality of service (QoS) to both protocols, to define and apply filters on the CM and CMTS, and to monitor traffic flowing through the cable network.

To support IPv6 multicast applications across DOCSIS cable networks, the CM and CMTS need to support multicast listener discovery version 2 (MLDv2) snooping. MLDv2 is identical to IGMPv3 and also supports ASM (any source multicast) and SSM (source-specific multicast) service models. Implementation effort on the CM/CMTS should be minimal because the only significant difference between IPv4 IGMPv3 and IPv6 MLDv2 is the longer addresses in the protocol.

IPv6 Addressing for Host, GWR, and ER   The host behind the CM will get a /64 prefix via stateless address autoconfiguration (SLAAC) or DHCPv6. If using SLAAC, the host listens for router advertisements (RAs) from the ER. The RAs contain the /64 prefix assigned to the segment. Upon receipt of an RA, the host constructs its IPv6 address by combining the prefix received in the RA (/64) and a unique identifier (e.g., its modified EUI-64 format interface ID).

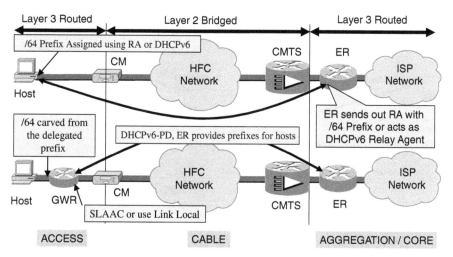

FIGURE 3.3 IPv6 addressing in a bridged environment.

If DHCPv6 is used to obtain an IPv6 address, it will work in much the same way as DHCPv4 works today. The DHCPv6 messages exchanged between the host and the DHCPv6 server are bridged by the CM and the CMTS. The GWR can use SLAAC to obtain an IPv6 address for its upstream interface, on the link between itself and the ER. This step is followed by a request via DHCP-PD (prefix delegation) for a prefix shorter than /64, which in turn is divided into /64 s and assigned to its downstream interfaces connecting to the hosts. The GWR sends out RAs on each of its downstream interfaces so that hosts connected to those interfaces can assign themselves an IPv6 address using SLAAC and the /64 prefix sent by the GWR.

The ER Layer 3 interfaces connected to the cable network can be configured manually with one or multiple /64 prefixes. Depending on the MSO's policy, the ER can be configured to send out RA messages downstream advertising a /64 prefix which can be used by the host and the GWR to assign an IPv6 address using SLAAC. If stateful DHCPv6 is used for address assignment, the ER can act as a DHCPv6 relay agent and forward DHCPv6 messages between the host/GWR and the DHCPv6 server. Figure 3.3 illustrates IPv6 addressing in a bridged CMTS environment.

IPv6 Routing in a Bridged CMTS Environment   The hosts install a default route that points to the ER or the GWR. No routing protocols are needed on these devices, which generally have limited resources. If a GWR is present, it will also use a static default route to the ER. The ER runs an IGP such as OSPFv3 or IS-IS for IPv6. The connected prefixes have to be redistributed. If DHCP-PD is used, a static route is installed by the ER with every prefix that is delegated. For this reason, the static routes must also be redistributed. Prefix

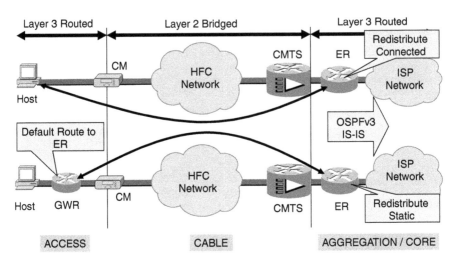

FIGURE 3.4 IPv6 routing in a bridged CMTS network.

summarization should be done at the ER. Figure 3.4 illustrates how IPv6 routing would work in a bridged CMTS environment.

3.2.2 Routed CMTS Networks

In a routed CMTS network the CM still acts as a Layer 2 device and bridges all data traffic between its Ethernet interface and the cable interface connected to the cable operator network. The CMTS acts as a Layer 3 router and may include the ER functionality. The hosts and the GWR use the CMTS as their Layer 3 next hop. When deploying IPv6, the CMTS/ER will need to either tunnel IPv6 traffic or natively support IPv6. To natively support IPv6, the CM and CMTS would need to be upgraded to DOCSIS 3.0. With pre-DOCSIS 3.0 cable networks (DOCSIS 1.0, 1.1, and 2.0) the only option for deploying IPv6 is tunneling. In the following sections are discuss different IPv6 deployment options in a routed CMTS network.

3.2.2.1 Deploying IPv6 with DOCSIS 2.0 DOCSIS 2.0-enabled CM and CMTS cannot natively support IPv6, due to the lack of support for neighbor discovery (ND) messages, which use multicast and MLDv2 snooping. The only way to deploy IPv6 is to tunnel it over the IPv4 infrastructure.

Figure 3.5 illustrates how IPv6 can be deployed in a DOCSIS 2.0 environment. In the first deployment scenario the host and ER are upgraded to dual-stack functionality, while the CM and CMTS remain IPv4. The IPv6 traffic is tunneled between the host and the ER using different IPv6 in IPv4 tunneling techniques. The tunnel source is the IPv4 address of the host, and the tunnel destination is an IPv4 address of the ER. The amount of configuration needed

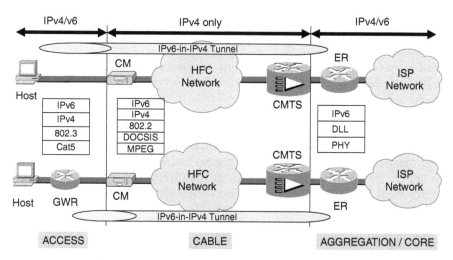

FIGURE 3.5 IPv6 deployment with DOCSIS 2.0.

on the host and ER to establish the tunnel will depend on the tunneling technique deployed. If the ER is terminating multiple tunnels, it would be advisable to use an automatic tunneling technique instead of manual tunneling, which may require more configuration. An advantage of using automatic tunneling on the ER is that the remote endpoints of the tunnels do not need to be specified manually, so all tunnel endpoints can be terminated on a single tunnel interface. This makes configuration on the ER simpler and easier to manage but may affect performance if a large number of tunnels are terminated on the same ER. For large-scale deployment it would be advisable to distribute the tunnels across multiple ERs. If IPv6 services such as multicast are being provided to the CPE, manual tunnels would need to be used, as automatic tunnels may not support IPv6 multicast. For IPv4, the host will use the CMTS as its Layer 3 next hop and forward all data traffic to it. The CM will remain a Layer 2 bridge and will follow the data-forwarding rules as outlined in the DOCSIS 2.0 specification.

In the second deployment model the host, GWR, and ER are upgraded to dual-stack functionality to support IPv6. The CM and CMTS remain as IPv4 devices. In this scenario, IPv6 in a IPv4 tunnel is established between the GWR and the ER. The GWR interface connected to the CM remains IPv4, while the interface connected to the host can be configured as a dual-stack functionality. Similarly, the ER interface facing the CMTS remains IPv4, while the interfaces connected to the ISP core/aggregation network can be configured in dual-stack mode. The host uses the GWR as the Layer 3 next hop for IPv4 as well as IPv6 traffic. The GWR uses the CMTS as its Layer 3 next hop for IPv4 traffic and the ER for IPv6 traffic.

The IPv6 traffic can be carried transparently over the HFC infrastructure as the CM and CMTS forward this traffic like regular IPv4 packets. One of the challenges with deploying IPv6 in the scenarios above is that it is very difficult to provide QoS to IPv6 traffic over the cable network. Since all IPv6 traffic is encapsulated in IPv4 packets, it is not possible to separate best-effort data traffic from real-time VoIP traffic needing a higher class of service. Therefore, all IPv6 traffic may receive the same class of service over the HFC infra-structure. It may be possible to mark or classify the IPv4-encapsulated IPv6 traffic at the tunnel endpoints (e.g., at the host/GWR or the ER), but this may not guarantee QoS to time/delay-sensitive IPv6 traffic over the cable network.

The IPv6 traffic can still receive a different class of service from that of IPv4 using the marking/classification at the tunnel endpoints; the CM and CMTS can use the marking/classification to direct IPv6 traffic onto service flows that get treatment that differs from best-effort service flows carrying IPv4 traffic.

IPv6 Addressing for Host, GWR, and ER   The host can get an IPv6 address once the tunnel interface comes up with the ER. The host can either use the /64 prefix it receives from the ER to configure itself with an IPv6 address using SLAAC or it can get its IPv6 address via DHCPv6. If the host is connected to a GWR, it can use the RA message sent by the GWR to assign itself an IPv6 address via SLAAC. For its IPv4 address the host can continue using DHCPv4, as the CM and CMTS will forward these IPv4 packets between the host and the DHCPv4 server.

The GWR can either use its link-local address for its upstream interface to communicate with the ER or use the RA messages sent by the ER to configure an IPv6 address on its upstream interface using SLAAC. The GWR can then act as a DHCP-PD requesting router and get a prefix delegated from the DHCPv6 server or the ER using DHCPv6 option 25 (OPTION_IA_PD) (RFC3633). Upon receiving the prefix delegated, the GWR can divide the prefix into /64 blocks and assign them to its downstream interfaces. The GWR can send out RA messages (the A-bit = 1) on these downstream interfaces so that the connected hosts can configure themselves with an IPv6 address using SLAAC. The GWR can also act as a lightweight DHCPv6 server and provide hosts with other network parameters, such as DNS and TFTP server addresses if the hosts request this information. The hosts can also obtain these network parameters from the MSO's DHCPv6 server. For its upstream interface IPv4 address and other network parameters, the GWR can continue to use DHCPv4 as it does today. For the downstream interfaces the GWR can act as a lightweight DHCPv4 server and assign IPv4 addresses to hosts from a configured pool. For IPv4 the GWR typically gets a public IP address on its upstream interface, but the downstream interfaces are assigned an RFC1918 private address. The IP addresses assigned to hosts are also private; therefore, the GWR implements a NAT functionality to translate between the public and private IP address space.

The ER is typically manually configured with a /64 IPv6 prefix that is included in the RA messages sent downstream. These RA messages are used by the host and GWR for assigning an IPv6 address to themselves. The ER can

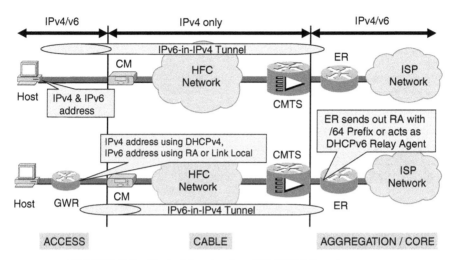

FIGURE 3.6   IPv6 addressing in a DOCSIS 2.0 environment.

also be configured to act as a DHCP-PD delegating router and to hand out
IPv6 prefixes to requesting routers. Figure 3.6 illustrates how IPv6 address
assignment is performed in a DOCSIS 2.0 environment.

IPv6 Routing in a DOCSIS 2.0 Environment   Depending on the tunneling
technique used, additional configurations might be needed on the GWR and
the ER. If the ER is also providing a 6to4 relay service, a default route will need
to be added to the GWR, pointing to the ER, for all non-6to4 traffic. If using
manual tunneling, the GWR and ER can use either static routing or an IGP
such as RIPng (RFC2080) or OSPFv3 (RFC2740). The IGP updates can be
transported over a manual tunnel, which does not work when using 6to4
tunneling since it does not support multicast.

Customer routes can be carried to the ER using RIPng updates. The ER can
redistribute the RIPng routes into its IGP, such as OSPFv3 or ISIS for IPv6.
Prefix summarization should be done at the ER. If DHCP-PD is used for
address assignment, a static route is installed on the ER automatically for each
prefix delegated. The static routes need to be redistributed into the IGP at the
ER so that there is no need for a routing protocol between the ER and the
GWR. The ER runs an IPv6 IGP such as OSPFv3 or ISIS for IPv6 toward
the ISP core network. Figure 3.7 illustrates how IPv6 routing would work in a
DOCSIS 2.0 environment.

3.2.2.2   *Deploying IPv6 Using Layer 2 Virtual Private Networks or Layer 2 Tunneling*   IPv6 over cable can also be deployed using features such as
L2VPN, which can transport IPv6 traffic from one subscriber to another using
a Layer 2 virtual circuit. On the CMTS, each CM is mapped to the appropriate

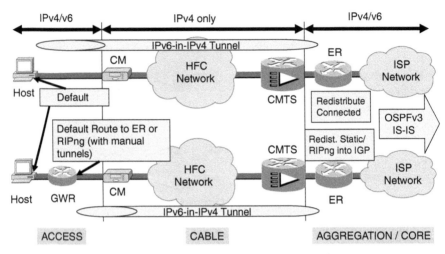

FIGURE 3.7   IPv6 routing in a DOCSIS 2.0 environment.

VLAN on the basis of its MAC address. The CMTS then creates an internal database of this one-to-one mapping of CM to VLAN, and uses it to encapsulate packets for the appropriate VLAN.

IPv6 traffic from CPE gets encapsulated in DOCSIS frames at the CM and is transported over the cable network. Once the traffic is received at the CMTS, it removes the DOCSIS header, adds a VLAN tag to the IPv6 packet, and forwards it to the aggregator. The aggregator performs inter-VLAN bridging or routing using integrated routing bridging (IRB), adds a new VLAN tag, and forwards packets to the appropriate CMTS. The CMTS then removes the VLAN tag, adds the DOCSIS header, and forwards the IPv6 packet downstream to the other CPE.

IPv6 traffic to and from a group of cable modems can be bridged into a single logical network by the aggregator by creating a secure virtual private network (VPN) for that particular group of cable modems. Traffic in one VLAN cannot be sent into another VLAN unless done specifically by the aggregator. Figure 3.8 illustrates how IPv6 traffic is transported over an L2VPN network.

When the CMTS receives a packet on an upstream, it looks up the service ID (SID) of the CM to see if it is mapped to a VLAN. If the mapping exists, and if the packet's source MAC address is not the CM's MAC address, the CMTS inserts the appropriate VLAN tag into the packet's header and forwards the packet out the appropriate WAN interface. If the packet is not being mapped, or if the packet originated from the CM, the CMTS forwards the packet using the normal Layer 3 routing.

When the CMTS receives a packet from a WAN interface (an interface that is connected to the aggregator or an upstream router) that is encapsulated with a VLAN tag, it looks in the internal database to see if the VLAN is

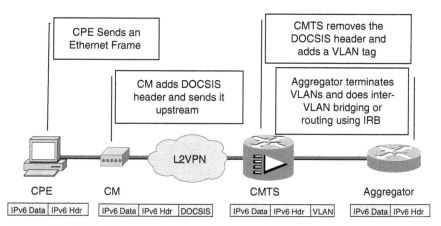

FIGURE 3.8 Transporting IPv6 traffic over an L2VPN network.

mapped to a CM. If so, the CMTS strips off the VLAN tag, adds the proper DOCSIS header, and transmits the packet on the appropriate downstream interface. If the packet is not being mapped, the CMTS continues with the normal Layer 3 processing of the packet. Since the IPv6 traffic is transported using Layer 2 tunnels or virtual circuits, there is no impact to the CMTS and CM. They can remain IPv4 devices and do not need to be upgraded to support IPv6. This type of deployment technique is used for providing IPv6 services to CPE devices only and cannot be used to provision and manage the CMTS and CM using IPv6.

IPv6 Addressing for Host and GWR The only devices in this deployment scenario that need to get an IPv6 address are the host and the GWR. Since IPv6 traffic is transported using Layer 2 tunnels, the CMTS and CM do not need an IPv6 address and continue to operate as IPv4-only devices. Figure 3.9 illustrates how the host and GWR are provisioned for IPv6 using L2VPN or Layer 2 tunnels. In the first scenario, PC-1 is connected directly to CM-1, which is a Layer 2 bridge and gets an IPv4 address using DHCPv4. In this case PC-1 can get an IPv6 address by using either DHCPv6 or SLAAC. In the second scenario, PC-2 is connected to the CM through a GWR. In this case, PC-2 can receive an IPv4 address from the GWR, which acts as a DHCPv4 server on its LAN interface connected to the PC. The GWR receives an IPv4 address on its interface connected to the CM via DHCPv4. The GWR uses either DHCPv6 or SLAAC to acquire an IPv6 address on its interface connected to the CM, followed by a DHCP-PD request. Once the GWR receives the IPv6 prefix that has been delegated, it chops the prefix into /64 chunks and sends the /64 prefix in its RA message on its LAN interface. PC-2, which is connected to the GWR LAN interface, uses SLAAC to configure itself with an IPv6 address using the /64 prefix received from the GWR.

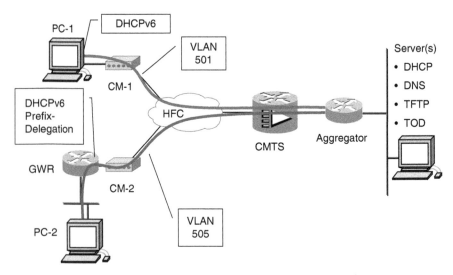

FIGURE 3.9   IPv6 addressing in a L2VPN network.

IPv6 Routing in a L2VPN Environment   All IPv6 traffic is transported over the L2VPN connection between the endpoints. The CMTS builds a virtual Layer 2 circuit between the CM and itself and forwards all traffic between the endpoints based on the static mapping configured for the endpoints. There is no need to enable any IPv6 routing protocols since all traffic is forwarded at Layer 2.

3.2.2.3 Deploying IPv6 with DOCSIS 3.0   DOCSIS 3.0 introduces full support for IPv6, including the provisioning and management of CM with an IPv6 address, and the ability to manage and transport IPv6 traffic over the cable infrastructure. With the changes made to the DOCSIS specification, the CM and CMTS are both IPv6 capable and can natively forward IPv6 traffic. To natively deploy IPv6 in a DOCSIS 3.0 environment, MSOs may need to upgrade hardware as well as software on their current CMTS and CM.

For initial deployments, IPv6 features can be supported in software since performance and scalability may not be a big concern. But as the IPv6 deployments begin to scale, the CMTS and CM would need to support IPv6 features in hardware in order to forward IPv6 traffic efficiently and not overburden the processing resources on these devices. It is important for the MSO to ensure that IPv6 services do not affect any existing IPv4 services in the network and provide users with the same or a better quality of experience as that associated with IPv4. This can definitely affect the rate of IPv6 acceptance in current SP networks.

Cable operators have several IPv6 deployment options in a DOCSIS 3.0 environment. They can upgrade the core and aggregation devices in the network to dual-stack status in order to support IPv6 and natively transport IPv6 traffic over the IP infrastructure. For MPLS-enabled MSO core, techniques such as MPLS 6PE/6VPE and EoMPLS could also be used to transport IPv6 packets. They can also upgrade the back-office servers to dual-stack status to manage and provision the CM, MTA, STB, and other end devices with IPv6. The edge of the network can either be dual-stack or IPv6-only.

If the lack of IPv4 address space is a key driver for the cable operator to deploy IPv6, it would make sense for the edge of the network to be IPv6-only. This would allow the cable operator to manage and provision the CM, MTA, STB, and other devices connected to the CMTS using IPv6. In this case the cable operator can continue to deploy new hardware and grow its subscriber base, which may not have been possible with IPv4, due to the shortage of addresses. The CMTS can be a dual-stack device using both IPv4 and IPv6 addresses, even when the devices connected to it are IPv6-only.

If the cable operator does not have IPv4 address constraints, the edge of the network can also be configured as a dual-stack functionality, which means that the CM, MTA, STB, and other devices connected to the CMTS can have an IPv4 as well as an IPv6 address. This can provide the cable operator with an opportunity to test with IPv6 in a production environment and ensure that current IPv4 services are not affected by enabling IPv6 in the network. Once cable operators are comfortable with IPv6, they can choose to turn off IPv4 at the edge of the network and configure the CM, MTA, STB, and other devices in IPv6-only mode. The cable operator can still run in dual-stack mode on the core and aggregation devices as well as the back-office servers, so they can continue to manage and support existing IPv4 devices in the network. If the MSO core network is MPLS-enabled, there is no need to upgrade the core to dual-stack status. In this case the core network is untouched and only the edge routers are upgraded to dual-stack mode to support MPLS-PE functionality.

There is a general consensus in the Internet community that IPv4 and IPv6 will continue to coexist in current networks for the foreseeable future, so certain portions of the network may be configured as dual-stack functionalities while the edge of the network may be configured as IPv6-only, for the reasons cited above.

DOCSIS 3.0 IPv6 Reference Architecture Figure 3.10 displays the DOCSIS 3.0 reference architecture and how IPv6 can be deployed in this environment. Three main DOCSIS 3.0 deployment models are represented in the figure:

1. The first model is the CM bridge model. In this deployment model the CM continues to be a Layer 2 bridge and forwards both IPv4 and IPv6 traffic. The CM can be provisioned and managed in IPv4-only, IPv6-only, or dual-stack mode. The CPE devices connected to the CM can still be provisioned via IPv4 and do not have to be upgraded to support IPv6. In this model the cable operator can benefit from the large IPv6 address space by provisioning and managing the CM exclusively with IPv6. This enables the cable operator to turn

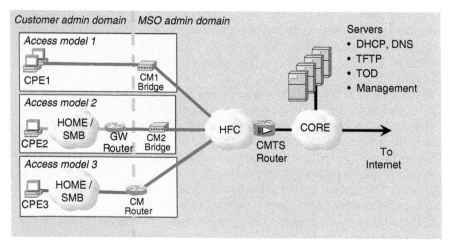

FIGURE 3.10   IPv6 deployment models in a DOCSIS 3.0 environment.

up new customers and continue to deploy new hardware without having to provide IPv6 services to customers. IPv6 can be deployed in a variety of phases using this approach. In this model the CM is provisioned and managed by the cable operator and the CPE devices connected to the CM are managed by the customer.

The CPE devices will use the CMTS as their next-hop gateway and will send all traffic directly to the CMTS. The CMTS will see the CPE devices as connected directly to its cable interface.

2. The second deployment model is the GW router model. In this deployment model the CM continues to be a Layer 2 bridge and forwards both IPv4 and IPv6 traffic natively in a manner similar to that for the first model. The CPE can be in IPv4-only, IPv6-only, or dual-stack mode. If the cable operator does not provide IPv6 services to the CPE, the customer can choose to deploy a GW router between the CM and the CPE, which can enable the customer to get IPv6 services and connectivity to the IPv6 Internet by tunneling IPv6 traffic over the cable operator's network. In this model the CM would be provisioned and managed by the cable operator, but the GW router and the CPE devices connected to it would be managed by the customer.

The CPE devices will use the GW router as the default gateway and will forward all traffic to it. The GW router will forward all traffic between the CMTS and the CPE devices. The CMTS will see the GW router as a device connected directly to its cable interface and will act as the Layer 3 next-hop router for the GW router.

3. The third model is the CM router model. In this deployment model the CM acts as a Layer 3 device and can have the GW router functionality built-in. The CM router can be provisioned and managed in dual-stack mode to forward both IPv4 and IPv6 traffic and provide both IPv4 end IPv6 services to the CPE

devices connected to it. In this model the CM router would be provisioned and managed by the cable operator.

The CPE devices will use the CM router as the Layer 3 next hop and will forward all traffic to it. The CM router will forward all traffic between the CPE devices and the CMTS. The CMTS will see the CM Router as a device connected directly to its cable interface and will act as a Layer 3 next hop for the CM router.

IP Provisioning for CM, GW Router, CM Router, and CPE   In a DOCSIS 3.0 environment the CM can perform IP provisioning in one of four modes: IPv4-only, IPv6-only, dual-stack mode, and alternate provisioning mode (APM). The CM can determine the IP provisioning mode by looking at type length value (TLV) 5 in the MAC domain descriptor (MDD) message sent down by the CMTS. In the absence of an MDD message the CM must use the IPv4-only provisioning mode. Figure 3.11 illustrates the format of the MDD message.

A DOCSIS 3.0-capable CMTS must transmit an MDD message periodically on every downstream channel in the media access control (MAC) domain. A MAC domain is defined as a downstream cable interface and its associated upstream interfaces. The CMTS MUST generate the MDD message in the format shown in Figure 3.11, including the parameters defined below.

- *Configuration Change Count* The CMTS increments this field by 1 whenever any of the values in this message change relative to the values in the previous MDD message sent on this downstream channel.
- *Number of Fragments* Fragmentation allows the MDD TLV parameters to be spread across more than one DOCSIS MAC frame. The value of this field represents the number of MDD MAC management frames that a unique and complete set of MDD TLV parameters are spread across to constitute the MDD message. This field is an 8-bit unsigned integer.
- *Fragment Sequence Number* This field indicates the position of this fragment in the sequence that constitutes the complete MDD message. Fragment sequence numbers start with a value of 1 and increase by 1 for each fragment in the sequence. Thus, the first MDD message fragment has

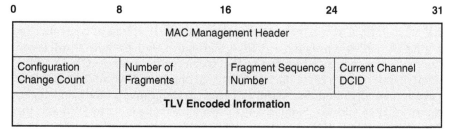

FIGURE 3.11   MAC domain descriptor (MDD) message format.

a fragment sequence number of 1, and the last MDD message fragment has a fragment sequence number equal to the number of fragments. This field is an 8-bit unsigned integer.

- *Current Channel Downstream Channel ID (DCID)* This is the identifier of the downstream channel on which this message is being transmitted.

All other parameters are encoded as TLV tuples, where the type and length fields are one octet each. A list of TLV parameters in the MDD message is included in Section 6.4.28 of the DOCSIS 3.0 specification. The TLV that communicates to the CM certain parameters related to the initialization of CM's IP-layer services is TLV 5. Here are the values that can be configured for TLV 5:

- 5.1: IP Version
  - 0 = IPv4
  - 1 = IPv6
- 5.2: IP Management Mode
  - 0 = single-stack operation
  - 1 = dual-stack operation
- 5.3: Alternate Provisioning Mode
  - 0 = If IP failure occurs, restart scanning of downstream channels
  - 1 = If IP failure occurs, fallback to nonpreferred IP version

The CM can be provisioned to operate in single- or dual-stack mode by setting the appropriate value in TLV 5.2. If the CM is provisioned to operate in single-stack mode, it can get either an IPv4 or an IPv6 address, but not both, as controlled by TLV 5.1. If the CM is provisioned to operate in dual-stack mode, it will try to acquire an IPv4 or an IPv6 address as controlled by TLV 5.1 and register with the CMTS. After registration the CM will try to acquire the second IP address type. For example, if TLV 5.1 is set to 0, the CM will try to acquire an IPv4 address and register with the CMTS. After registration the CM will try to acquire an IPv6 address. But if TLV 5.1 is set to 1, the CM will try to acquire an IPv6 address and register with the CMTS. After registration the CM will try to acquire an IPv4 address. In both cases, if the CM fails to acquire the second IP address, it will not go off-line.

If the CM is provisioned to operate in APM mode as controlled by setting TLV 5.3 to 1, the CM will try to acquire the primary IP address as controlled by TLV 5.1 (0 = IPv4 is primary, 1 = IPv6 is primary), and if it fails, it will try to acquire the nonprimary IP address so that it can register with the CMTS. If the primary mode is set to IPv6, the CM will try to obtain an IPv6 address; if it fails to acquire an IPv6 address, it will fall back to nonprimary mode and try to acquire an IPv4 address. Similarly, if the primary mode is set to IPv4 and the CM fails to acquire an IPv4 address, it will fall back to nonprimary mode and try to acquire an IPv6 address.

Regardless of which mode the CM is provisioned in, it will probably use DHCP to acquire an IP address. If the CM is trying to acquire an IPv4 address, it will initiate a DHCPv4 handshake with the DHCP server, and if the CM is trying to acquire an IPv6 address, it will initiate a DHCPv6 handshake with the DHCP server. The CMTS will act as a DHCP relay in both cases and forward the DHCP messages between the CM and the DHCP server.

In the first deployment model illustrated in Figure 3.10, the CM can use DHCPv6 to acquire an IPv6 address or DHCPv4 to acquire an IPv4 address. The CPE connected to the CM can also use DHCPv4 or DHCPv6 to acquire its IP address from the cable operator's DHCP server.

In the second deployment model illustrated in Figure 3.10, the CM and GW router can use DHCPv4 or DHCPv6 or both to acquire their IP addresses. From the CMTS's perspective, the GW router appears as a CPE device connected to the CM, so the GW router may receive only a single IPv4 address or a single IPv6 prefix. The host connected to the GW router will need to acquire its IP address from the GW router, which would need to be configured as a DHCP server on the interface connected to the host. For IPv4, the GW router may also need to perform network address translation (NAT) between its interface connected to the host (the inside interface) and the interface connected to the CM (the outside interface). For IPv6, GWR can act as a DHCP-PD requesting router and receive a delegated prefix from the DHCPv6 server. It can also send out RA messages to the CPE devices connected to it, advertising the IPv6 prefix to be used by CPE devices for IPv6 address assignment using SLAAC.

In the third deployment mode illustrated in Figure 3.10, the CM router can use DHCPv4 or DHCPv6 to acquire its IP address. If the CM router acquires an IPv4 address, it will need to act as a DHCPv4 server for the CPE connected to it and provide NAT functionality as described above. If the CM router uses DHCPv6, it can receive an IPv6 prefix assigned to its interface connected to the CMTS and then send a DHCP-PD request to the server in order to get a delegated prefix. This delegated prefix can be carved into several /64 prefixes and advertised on the CM router's downstream interfaces, to which CPE devices are connected. The CM router can include the /64 prefixes in the router advertisement messages sent out periodically on its downstream interfaces. The CM router will also need to set the A-bit = 1 and O-bit = 1, which will instruct the CPE devices to use SLAAC to configure themselves with an IPv6 address and then request other network parameters from the DHCP server.

Refer to Chapter 2 for a detailed explanation of the use of A, M, and O bits. These bits are used to control the behavior of hosts for using either stateful DHCPv6 or SLAAC and stateless DHCPv6 for IPv6 address assignment and obtaining other network parameters.

IPv6 Routing in a DOCSIS 3.0 Environment   In the first model illustrated in Figure 3.10, there is no need for a routing protocol between the CM and CMTS since the CM acts as a Layer 2 bridge and forwards all IPv4 as well as IPv6 traffic between the CPE and the CMTS. The CPE will have a default route (both for IPv4 and IPv6) pointing to the CMTS as the Layer 3 next hop.

In the second model illustrated in Figure 3.10, the GW router can use a static default route pointing to the CMTS as the Layer 3 next hop. Since the GW router is not controlled by the MSO, there may not be a need to run any routing protocol between the GW router and the CMTS.

In the third model illustrated in Figure 3.10, the CM router can use a static default route pointing to the CMTS, or it can run a routing protocol such as RIPng or OSPFv3 between itself and the CMTS. Customer routes from behind the CM router can be carried to the CMTS using dynamic routing updates.

If DHCP-PD is used for address assignment, a static route is installed automatically on the CMTS for each delegated IPv6 prefix. The static routes need to be redistributed into the IPv6 IGP at the CMTS so that there is no need for a routing protocol between the CMTS and the CM router. The CMTS can run an IGP such as OSPFv3 or ISIS for IPv6 on its interfaces that are connected to the IP Core network.

3.2.2.4 Other Considerations for Deploying IPv6 in Cable (DOCSIS) Networks The cable operator may face several challenges when deploying IPv6 over its cable infrastructure. These challenges include providing the appropriate quality of service (QoS) to IPv4 and IPv6 traffic, security considerations for IPv4 as well as IPv6 traffic, and network management–related issues. Some of these challenges are discussed briefly below.

IPv6 QoS In a DOCSIS environment the traffic between the CMTS and CM is carried over unidirectional service flows. A service flow is defined as a MAC-layer transport service that carries traffic in the upstream and downstream directions. To provide the appropriate QoS to IPv6 traffic, the CM and CMTS would need to classify the IPv6 traffic properly using different classifiers and then direct this traffic to appropriate service flows.

A classifier is defined as a set of matching criteria applied to each packet entering the DOCSIS network which consists of some packet-matching criteria (destination IP address, traffic priority, IP precedence, etc.) and a classifier priority. A QoS classifier additionally consists of a reference to a service flow. If a packet matches the packet-matching criteria specified for a QoS classifier, it is then delivered on the referenced service flow. The service flows can be configured with a variety of QoS parameters, such as guaranteed reserved bandwidth, high traffic priority, and higher burst size.

By using classifiers the IPv6 traffic can be transported over the DOCSIS network with a higher QoS than that for other best-effort traffic. If the IPv6 traffic cannot be classified using the defined set of matching criteria, it will typically be carried over default service flows, which are treated as best effort and do not receive any special QoS treatment.

The DOCSIS 3.0 specification provides for a QoS framework for IPv6 as well as IPv4 and allows traffic to be classified using a defined set of matching criteria for transport over the DOCSIS network with the appropriate QoS handling. For example, IPv6 time-sensitive traffic such as voice and video can be carried over high-priority queues in the upstream and downstream directions using the appropriate classifiers at the CM and CMTS. The regular data traffic can be carried over default best-effort service flows.

IPv6 Security   Security in a DOCSIS cable network is provided using baseline privacy plus (BPI+), which covers CM authentication, key exchange, and establishing encrypted traffic sessions between the CM and CMTS. The only part that is dependent on IP addresses is encrypted multicast. Semantically, multicast encryption would work the same way in an IPv6 environment as in the IPv4 network. DOCSIS uses public-key cryptography to establish a shared secret (an authorization key) between the CM and its CMTS. The shared secret is then used to derive secondary keys, which in turn are used to secure subsequent baseline privacy key management (BPKM) exchanges of traffic encryption keys. This two-tiered mechanism for key distribution permits traffic encryption keys to be updated without incurring the overhead of computationally intensive public-key operations.

A CMTS authenticates a client CM during the initial authorization exchange, which occurs when early authentication and encryption (EAE) is enabled by setting TLV 6 to 1 in the MDD message sent by the CMTS, or when postregistration BPI + is enabled by setting TLV 29 to 1 in the CM DOCSIS configuration file. Each CM carries a unique digital certificate issued by the CM's manufacturer. The digital certificate contains the CM's public key, along with the CM MAC address, the identity of the manufacturer, and the CM serial number. When requesting an authorization key, a CM presents its digital certificate to a CMTS. The CMTS verifies the digital certificate and then uses the CM's public key to encrypt an authorization key, which the CMTS sends to the requesting CM.

For hosts to enhance security, limited changes have to be made. Privacy extensions (RFC3041) for SLAAC should be used by the hosts. IPv6 If available on the host or the GWR, firewall functions could be enabled. The cable operator provides security against attacks that come from its own subscribers, but it can also implement security services that protect its subscribers from attacks from sources external to the network.

The CMTS should protect the cable operator's network and the other subscribers against attacks by one of its own customers. For this reason, unicast reverse path forwarding (RFC3704) and ACLs should be used on all interfaces facing subscribers. Filtering should be implemented with regard for the operational requirements of IPv6. The CMTS should protect its processing resources against floods of valid customer control traffic, such as router and neighbor solicitations, router and neighbor advertisements, duplicate address detection (DAD) requests, MLD requests, and so on. The CMTS can also use the secure neighbor discovery (SEND) (RFC3971) protocol to secure neighbor discovery traffic.

All other security features used with the IPv4 service should be applied similarly to IPv6. In addition to the above-mentioned security measures, the DOCSIS 3.0 security specification introduces the following new features:

- *Early authentication and encryption (EAE)* Early authentication functions as a network admission control; only authenticated CMs are allowed to continue their initialization process and may subsequently be admitted to the network. The results of a successful authentication are used to secure subsequent steps in the CM's initialization process. EAE includes authentication of the CM after it completes the ranging process and before it initiates the DHCP message exchange. It also includes exchanging traffic encryption key (TEK) key for the CM and encryption of IP provisioning traffic and the registration request MAC message during CM initialization.

- *Secure provisioning* The term secure provisioning refers to securing the CM provisioning processes. These processes include securing DHCP, ToD, and TFTP at the IP layer, and registration at the MAC layer. Secure provisioning plays a critical role in protecting the CMs and the network from attacks and in preventing service theft.

- *Securing DHCP messages* DHCP is a client–server protocol. When EAE is enabled for a CM, security of DHCP messages between the CMTS and the CM is provided by encrypting DHCP packets as they pass across the DOCSIS network. This helps in protecting the IP address and other network parameters assigned to the CM in the DHCP message exchange from hackers who may try to gain access to this information in order to launch a denial of service (DOS) attack against legitimate customers as well as the cable operator's network resources.

- *TFTP configuration file security* The CMTS supports several features that can be used to secure downloading of the CM DOCSIS configuration file and its contents. One of the features is protecting the actual IP address of the TFTP server, which stores the DOCSIS configuration file. The CMTS provides this feature by acting as a TFTP proxy and by intercepting TFTP requests from the CM. When a CM tries to download the DOCSIS configuration file from the TFTP server, the CMTS intercepts this requests, downloads the DOCSIS configuration file from the TFTP server on the CM's behalf, and sends it down to the CM. The CMTS can also encrypt the file name of the DOCSIS configuration file as well as its contents so that someone trying to sniff this traffic cannot gain access to the contents of the DOCSIS configuration file, which contains the class of service assigned to the CM. By doing this, the CMTS can prevent theft of service and also protect legitimate customers from being denied access to the network by hackers who can sniff the contents of the DOCSIS configuration file and gain access to network resources pretending to be a legitimate, paying customer.

- *Securing registration request (REG-REQ-MP) messages* When the CM completes downloading the DOCSIS configuration file, it send a registration request message to the CMTS and includes in this message, a subset of the parameters included in the configuration file. To maintain the confidentiality of these parameters, when EAE is enabled, the CM must encrypt the REG-REQ-MP message it sends to the CMTS.

- *Source address validation (SAV)* The CMTS is responsible only for forwarding CPE packets that contain legitimate addresses. When the SAV feature is enabled, the CMTS must drop any packets received upstream whose IP source address has not been assigned by the cable operator. This includes packets whose source IP address is an IP address that has been assigned to another device. Source IP addresses are considered assigned by the operator when they are provisioned via DHCP messaging or identified by parameters in the configuration file.

IPv6 Network Management IPv6 can have many applications in DOCSIS networks. Cable operators can initially implement IPv6 on the control plane and use it to manage the thousands of devices connected to the CMTS. This would be a good way to introduce IPv6 in a DOCSIS network. Later the cable operator can extend IPv6 to the data plane and use it to carry customer as well as management traffic. It is important to note that even when cable operators use IPv6 for management of large numbers of devices in the network, they can continue to use IPv4 transport and poll IPv6 MIB over IPv4.

IPv6 can be enabled in a DOCSIS network for management of devices such as CM, CMTS, CM router, and ER. With the roll-out of advanced services such as VoIP and video over IP, cable operators are looking for ways to manage the large number of devices connected to the CMTS. In IPv4, an RFC1918 address is assigned to these devices for management purposes. Since a finite number of RFC1918 addresses are available, it is becoming difficult for MSOs to manage these devices.

By using IPv6 for management purposes, cable operators can scale their network management systems to meet their needs. The CMTS can be configured with a /64 management prefix, which is shared among all CMs connected to the CMTS cable interface. Addressing for the CMs can be done via SLAAC or DHCPv6. Once the CMs receive a /64 prefix, they can configure themselves with an IPv6 address. If there are devices behind the CM that need to be managed by the cable operator, another /64 prefix can be defined on the CMTS. These devices can also use SLAAC or DHCPv6 to assign themselves an IPv6 address.

Traffic sourced from or destined to the management prefix should not cross a cable operator's network boundaries. In this scenario IPv6 will be used only for managing devices on the DOCSIS network. The CM will no longer require an IPv4 address for management, as described in earlier sections.

The current DOCSIS, PacketCable, and CableHome MIB (management information base) modules have been designed to support IPv6 objects. In this case, IPv6 neither adds nor changes any of the functionality of these MIB modules. The textual convention used to represent Structure of Management Information version 2 (SMIv2) objects representing IP addresses was updated (RFC4001) and a new Textual Convention InetAddressType was added to identify the type of IP address used for IP address objects in MIB modules.

There are some exceptions. The MIB modules required to add IPv6 support are defined in the DOCSIS 3.0 specification. In general, the CMTS and the CM must support the RFC3419 recommendations to support SNMP over IPv6.

3.3 SUMMARY

Several options are available to cable operators for deploying IPv6 over their IP and DOCSIS infrastructure. Depending on the current state of the network and the main drivers for deploying IPv6, the cable operator can choose the option that best suits its needs. The ideal scenario is to deploy IPv6 natively and upgrade the CM, CMTS, back-office servers, and core (except in the case of MPLS-based core) devices in the network to transport IPv6 traffic natively across the network. This will allow the cable operator to deploy IPv6-based value-added services such as IP multicast voice and video applications for its customer base.

Deploying IPv6 natively in the network can be costly at first since a large part of the network may have to be upgraded to support IPv6. But the long-term benefits can offset the upfront cost since the cable operator can design an infrastructure that can last for several years and that will meet the scalability and performance requirements for its growing customer base.

REFERENCES

1. CableLabs, "MAC and Upper Layer Protocols Interface Specification (CM-SP-MULPIv3.0-I07-080215)," Data-Over-Cable Service Interface Specification (DOCSIS) 3.0, February 15, 2008.
2. CableLabs, "Security Specification (CM-SP-SECv3.0-I07-080215)," Data-Over-Cable Service Interface Specification (DOCSIS) 3.0, February 15, 2008.
3. CableLabs, "Operations Support System Interface Specification (CM-SP-OSSIv3.0-I06-080215)," Data-Over-Cable Service Interface Specification (DOCSIS) 3.0, February 15, 2008.
4. Y. Rekhter, R. Moskowitz, D. Karrenberg, G. Groot and E. Lear, "Address Allocation for Private Internets," BCP 5, RFC1918, February 1996.
5. G. Malkin and R. Minnear, "RIPng for IPv6," RFC2080, January 1997.
6. T. Narten and R. Draves, "Privacy Extensions for Stateless Address Autoconfiguration in IPv6," RFC3041, January 2001.
7. IAB and IESG, "IAB/IESG Recommendations on IPv6 Address Allocations to Sites," RFC3177, September 2001.

8. M. Daniele and J. Schoenwaelder, "Textual Conventions for Transport Addresses," RFC3419, December 2002.

9. O. Troan and R. Droms, "IPv6 Prefix Options for Dynamic Host Configuration Protocol (DHCP) Version 6," RFC3633, December 2003.

10. F. Baker and P. Savola, "Ingress Filtering for Multihomed Networks," BCP84, RFC3704, March 2004.

11. M. Daniele, B. Haberman, S. Routhier and J. Schoenwaelder, "Textual Conventions for Internet Network Addresses," RFC4001, February 2005.

12. J. Arkko, J. Kempf, B. Zill and P. Nikander, "Secure Neighbor Discovery (SEND)", RFC 3971, March 2005.

13. R. Coltun, D. Ferguson and J. Moy, "OSPF for IPv6," RFC2740, December 1999.

4 IPv6 Deployment in DSL, ETTH, and Wireless Networks

In this chapter we discuss deploying IPv6 in DSL, ETTH, and wireless networks. Various deployment models, along with the relevant network components, are described in detail. We also cover some of the challenges faced by service providers (SPs) when deploying IPv6 in their current environments.

4.1 NEW REMOTE ACCESS ARCHITECTURE FOR IPv6

IPv6 was designed to provide incremental benefits over IPv4. Its main features were driven primarily by an anticipated shortage of IPv4 addresses—thus the larger IPv6 address space and incremental improvements aimed at addressing shortcomings of IPv4. Therefore, in many aspects of networking, such as routing, the differences between IPv4 and IPv6 are minimal and IPv6 does not present significant architectural changes over IPv4. The techniques associated with remote access deployment are tightly coupled with the fundamentals of IP. Remote access makes extended use of Layer2–Layer3 mapping protocols, prefix allocation, PPP (point-to-point protocol) interaction, autoconfiguration, and other factors. Therefore, innovations and philosophical changes introduced by IPv6 tend to affect remote access techniques that are more dependent than other areas of networking. In the following sections we highlight some of the main differences in routing, address allocation, subscriber management, and other areas.

4.2 DSL NETWORKS

Digital subscriber line (DSL) is the dominant access technology deployed by most large networks in Europe and Japan. The popularity of DSL stems from the ease of deployment, which leverages the physical infrastructure of

Deploying IPv6 in Broadband Access Networks, By Adeel Ahmed and Salman Asadullah
Copyright © 2009 John Wiley & Sons, Inc.

the existing telephone network. The bandwidth available to subscribers depends on the quality of the telephone twisted pair as well as on the distance between the subscriber and the first active equipment, referred to as the DSL access multiplexer (DSLAM). The average bandwidth available through this technology makes it appropriate to support the applications and network deployments envisioned with IPv6. Although it is theoretically possible to offer IPv6 over a 56 K modem access line, it is almost certain that only one host could be supported on such a low-speed line. This will not justify the cost involved in deploying IPv6, which offers static and large address allocations.

As mentioned in Chapter 1, large-scale commercial IPv6 access deployment was begun around 2003 by NTT Japan. Encouraged by government incentives, NTT deployed an IPv6 DSL network as documented in RFC4241. This deployment leveraged existing DSL technology with PPP as a Layer2 protocol, thus employing a common IPv4 deployment technique.

4.2.1   DSL Network Elements

DSL access solutions are widely deployed. However, design complexities and diversities present significant challenges for migrating to IPv6. A number of base elements went through the standardization process, mainly in the IETF. Deployment techniques come in varying flavors, with some requiring components of the infrastructure to be IPv6-enabled in a mandatory fashion while others suggest a more relaxed transition path from IPv4, with permissible exchange of control messages over IPv4 transport. Layer2 encapsulations such as ATM and PPP need to support IPv6 as a Layer3 protocol. DHCP and RADIUS are still used as the main configuration and provisioning protocols. They have both been modified to support IPv6, but they continue to retain their key roles in the new deployment models.

A typical DSL deployment model is depicted in Figure 4.1. It includes scenarios in which a network access provider (NAP) offers transport facility for an internet service provider (ISP) or provides IPv6 connectivity directly.

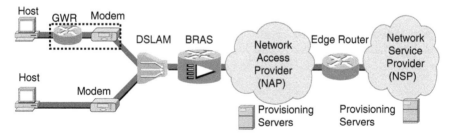

FIGURE 4.1   Typical DSL deployment model.

To deploy IPv6 service in this environment a number of network elements need to be upgraded.

- *Host.* The host needs to run an IPv6 stack. In many cases the host is connected to the IPv6 network directly through a Layer2 Ethernet connection. In other cases, a PPP session is established between the host and the broadband remote access server (BRAS), and possibly relayed to an edge router (ER) for termination. In this case, the host runs a PPPoE client that supports IPv6 traffic.

- *GWR.* In most cases, the Layer3 gateway router (GWR) runs an IPv6 stack to support host connectivity. When the GWR cannot be upgraded, the host tunnels IPv6 over IPv4 or over PPP, and the GWR can stay in IPv4-only mode.

- *Modem.* The modem acts as a Layer2-only device and needs to be transparent to network protocols, including IPv6. As the modem offers a downstream Ethernet interface, no restrictions apply to this Ethernet interface related to upper layer protocols.

- *DSLAM.* The digital subscriber line access multiplexer (DSLAM) is a network device, usually at a telephone company central office, that receives signals from multiple customer digital subscriber line (DSL) connections and multiplexes the signals onto a high-speed backbone link.

- *BRAS.* The broadband remote access server (BRAS) is enabled for IPv6 when PPP sessions are terminated on it. Otherwise, it tunnels PPP sessions over L2TP (the Layer2 tunneling protocol) and does not need to support IPv6. The connection from the BRAS to the RADIUS server, provided by the network access provider (NAP), is established over IPv4 transport.

- *NAP.* The network access provider (NAP) can use IPv4 transport when relaying PPP sessions between the BRAS and the edge network. L2TP is capable of using IPv4 as a transport protocol; migration toward IPv6 is not mandatory at this stage. If the BRAS terminates the PPP sessions, it will need to support IPv6 to offer connectivity.

- *NSP.* The network service provider (NSP) can terminate the PPP and L2TP sessions forwarded by the BRAS. The NSP network needs to be IPv6-aware in order to provision IPv6 users as well as to provide Layer3 services to them.

- *ER.* The edge router (ER), which is part of the NSP network, is IPv6 enabled in all deployment scenarios.

- *Provisioning servers.* These include the AAA RADIUS server for authenticating and authorizing users and an DHCP server for IPv6 address assignment. The DHCP server is used for IPv6 address assignment as well as to provide other network parameters to clients. DHCPv6 functionality is also required for IPv6 deployment, as PPP and RADIUS alone cannot fully provision an IPv6 subscriber. The DNS server is used for name-resolution purposes; the same DNS server can be used for IPv4 and IPv6.

4.2.2   DSL Service Models

In Chapter 1 we described the two architectures that are widely deployed in DSL environments: the ISP-operated deployment model, without L2TP; and the wholesale deployment model, based on L2TP. These architectures build on the capability of PPP to create virtual circuits over an infrastructure that always appealed to SP because it allows control of user traffic by taking advantage of the traffic engineering capabilities of the underlying ATM protocol.

Both models have a significant impact on such issues as subscriber management, routing, and address allocations. This is caused by new approaches introduced by DHCP and PPP for IPv6 support. Both models are PPP-based. Being the most common approach to offering IPv4 Internet service, PPP was quickly adopted by many SPs for initial IPv6 service offerings. Connecting users to the Internet with dedicated Ethernet (VLAN) or WiFi links often lead to non-PPP-based deployment models. A number of SPs are currently exploring new IP-based (independent of IP version) deployment models, most of which are still undergoing standardization and are therefore not on a par with PPP-based models in terms of technology maturity.

4.2.2.1 ISP-Operated Deployment Model   The ISP-operated deployment model is illustrated in Figure 4.2. In this broadband access architecture, the BRAS terminates the PPP session. The subscriber is able to run a PPPoE client to terminate the PPP session when the GWR is a Layer2 device bridging between Ethernet and ATM media. The RADIUS dialog between the BRAS and the NAP RADIUS server occurs over an IPv4 transport.

When IPv6 routing is not suitable to direct traffic to the ISPs, a VLAN or any other form of tunnel can be created on a per ISP basis from the BRAS to the ER, providing routing isolation between the NAP and ISP routing domains. Terminating PPP sessions on the BRAS yields some advantages and disadvantages.

- *Advantages*
  - The traffic enters the IP routing domain as soon as the PPP sessions are terminated on the BRAS, which results in routing optimization of unicast and multicast packets. For unicast packets, it avoids the use of a suboptimal path that leads to provisioning of unused capacity in the

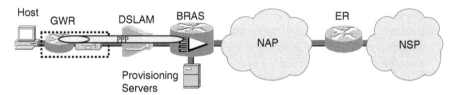

FIGURE 4.2   ISP-operated deployment model.

backbone. For multicast packets, this model provides the replication point close to the subscriber, resulting in conservation of the NAP resources. With a NAP infrastructure-supporting multicast, each multicast flow will appear only once per BRAS with a registered subscriber.

- Multicast is a popular IPv6 application, and its deployment is more scalable, as group allocation is made simpler by the larger IPv6 address space. Among other advantages, multicast group numbers can be derived from the IPv6 unicast allocation. In contrast, the L2TP model forces the multicast traffic to be replicated at the ER, overloading the NAP infrastructure with multicast replications.

- *Disadvantage*
    - A key disadvantage related to this model is that the NAP infrastructure should have full IPv6 support, as it provides Layer3 services between the BRAS and the ER. This approach makes the IPv6 network dependent on multiple entities with regard to resources for learning, designing, deploying, and support operations. These requirements also apply to the NAP RADIUS server, which needs to support IPv6 attributes to manage and provision IPv6 subscribers.

4.2.2.2 Wholesale Deployment Model Figure 4.3 depicts a wholesale architecture with L2TP access concentrator (LAC) and L2TP network server (LNS) elements. The ISP subscriber virtual link supports IPv6 in order to provide an IPv6 service to the end user. L2TP, the tunneling mechanism between the LAC, and the LNS, is operated over IPv4. Similarly, the RADIUS dialog between the BRAS and NAP RADIUS server, and the edge router and ISP RADIUS server, is performed over an IPv4 transport as well. In this model, the subscribers' database is managed by the ISP RADIUS server. Using L2TP in this model has a number of trade-offs:

- *Advantage*
    - IPv6 network support is required only on the host, the GWR, and the ER. The NAP infrastructure does not need to support IPv6. The BRAS continues to operate as an IPv4-only device. The SP providing this service may not have to interact with the NAP to deploy IPv6. This makes the deployment of IPv6 easier.

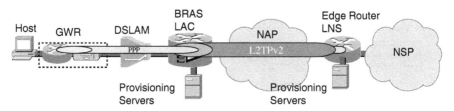

FIGURE 4.3 Wholesale deployment model.

- *Disadvantages*
  - PPP sessions are maintained at the BRAS and the ER, thus overutilizing the overall network resources. As IPv4 and IPv6 will continue to coexist in the network, the support of IPv6 adds cost at multiple locations. If the option to run a separate PPP session for IPv6 is chosen, both BRAS and ER are affected.
  - Since IPv6 multicast packets need to be replicated at the ER, the NAP infrastructure is overutilized, and the L2TP tunnels between ER and BRAS are overloaded.

4.2.2.3 Hybrid Model   In the scenarios depicted in earlier sections, a number of options are available, based primarily on the use of PPP tunnels to segregate IPv6 and IPv4 traffic and infrastructure. With the perceived immaturity of IPv6, segregated IPv4 and IPv6 deployment is seen as an appealing option. This infrastructure segregation can also be driven by technical considerations related to traffic engineering. One may also question the coexistence of IPv4 and IPv6 traffic, which have very different business impacts. IPv4 business is mature and revenue generating, and IPv6 applications are relatively new to the market.

For this deployment scenario, PPP offers an easy option (as it is based on virtual circuits) to push IPv4 and IPv6 traffic originated by the subscribers to different endpoints (Figure 4.4). Two PPP sessions, one for IPv4 and one for IPv6, are originated by the subscriber. Traffic separation is achieved by configuring only one network protocol per PPP session. If an L2TP-based model is deployed for IPv4 to offer connectivity to the ISP, the IPv6 service can be deployed in a way that does not interact with IPv4 service. The IPv6 PPP sessions are terminated on the BRAS. This enables the NAP to offer IPv6 connectivity that either does not require access to the Internet, or has a dedicated path to access the IPv6 Internet. IPv6 multicast would be one of the applications that can take advantage of this architecture. The NAP continues to offer transport service to ISPs but runs an IPv6 multicast service that does not require Internet access. Even though this model looks complex at first sight, it presents a number of advantages and disadvantages.

- *Advantages*
  - The business model based on L2TP is untouched and can keep operating independently.

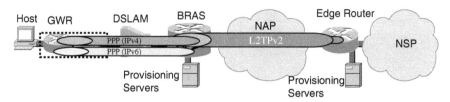

FIGURE 4.4   Hybrid access model for IPv4 and IPv6.

- The PPP sessions for both IPv4 and IPv6 are terminated on different devices, avoiding sharing between IPv6 and IPv4 traffic. This can help ensure that IPv6 is deployed in the network without affecting existing IPv4 services.

- *Disadvantages*
  - If both IPv4 and IPv6 Internet access is provided, troubleshooting connectivity problems becomes challenging. If the IPv4 and IPv6 DNS zones are disjoint, it is less of a problem, but if both IPv6 and IPv4 records are returned for the same name, it can lead to complications since the host can chose to establish a connection using either IPv4 or IPv6.
  - The BRAS is now fully IPv6 aware and needs to be upgraded. In the L2TP model, the BRAS is performing minimal IPv4 routing, as it just relays PPP sessions to tunnel endpoints. For IPv6, it is now a fully featured edge device with QoS, policing, and routing functions.
  - The number of PPP connections terminated on the BRAS is multiplied by a factor of 2, which makes this model unsuitable for large-scale deployment.

4.2.3 Protocol Considerations

So far, we have considered IPv6 deployment issues in DSL infrastructure options. Now let us delve into the details of how protocols such as PPP and DHCP, the two pillars of any access architecture, are affected by IPv6.

4.2.3.1 PPP Design Options   PPP supports multiple network protocols at the same time. By virtue of a layered approach, multiple Layer3 network protocols can run on the same PPP session. In the early days of the Internet, it was common to run non-IP protocols such as AppleTalk or IPX—ATCP (RFC1378) for AppleTalk and IPXCP (RFC1552) for IPX—on the same PPP session. In the context of low-speed dial-up networks, the only design choice was to run multiple network control protocols (NCPs) on the same PPP session (RFC1661). These days, IPv6 network designers are presented with more deployment options. The first approach is to run one PPP session and multiple NCPs: namely, IPCP and IPV6CP (RFC2472). This keeps the number of PPP sessions the same as in the current IPv4 environment. Another approach is to run one PPP session per Layer3 protocol—one for IPv4 and one for IPv6—even if those two sessions are terminated on the same gateway. The per protocol session approach has the advantage of keeping IPv6 on a separate session so that it does not affect the IPv4 session. This may facilitate debugging of pioneering IPv6 networks and presents a safer approach, as IPv4 will not be affected by an IPv6 session failure.

On the other hand, this dual-session approach puts pressure on the SP equipment, as the number of sessions is multiplied by a factor of 2 for each dual-stack subscriber. This may result in doubling SP resources allocated to PPP. Financial considerations may affect these types of decisions, as SPs are generally under pressure to lower their operational costs.

On the subscriber side of the PPP session, scalability may not be an issue as in the current IPv4 triple-play approach; it is common to have the subscriber equipment running a PPP session per service (e.g., data, voice, and video), each of these PPP sessions being terminated on different SP endpoints. It is likely that the introduction of IPv6 in this model is not going to double the number of PPP sessions. Although it is wise to offer a dual-stack Internet access, for closed services such as voice or video, the SP can deliver it over IPv4 or IPv6 transport without affecting either service. So in the best-case scenario, the introduction of IPv6 will simply add one additional PPP session for basic IPv6 connectivity.

4.2.3.2 PPP for IPv6   PPP plays a key role in a large number of remote access architectures. Although it was invented during the time of low-speed access technologies such as dial-up and ISDN, it is instrumental in offering Layer2 connectivity for point-to-point DSL deployments. This is a natural way of transporting IPv6 over PPP links (RFC2472). The RFC1332 (the PPP IPCP) defines a way to transport IPv4 optimally over PPP. It offers a number of negotiable options for Layer3 and above, such as IPv4 address, DNS server address, and mobile IP (RFC2290). This approach is based on interlayer dependency in negotiating upper layer protocol information using PPP. IPv6 over PPP departs from this approach.

IPv6 over PPP is defined in a way that builds on previous experiences and does not define any layer violations. The two options available in IPV6CP are the interface identifier and the IPv6 compression protocol. The interface identifier differs from an IPv6 address, as it includes only 64 bits instead of 128. In the PPP case, the interface identifier offers the remote party the final 64 bits of the IPv6 address. Combined with the prefix found in a router advertisement (RA) sent on the same link, the remote end of the PPP link can fully specify the IPv6 address of the other end. The notable difference is that full IPv6 address allocation is not possible with PPP only. It has to be completed by two protocols: neighbor discovery and IPV6CP or DHCPv6. The RA message is sent using ND, specifying the first 64 bits of the IPv6 address, and the interface identifier, defining the last 64 bits, is sent using IPV6CP.

Another key aspect of IPv6 is the large allocations of addresses to subscribers. Initially, PPP was used in providing IPv6-based services. However, with the maturity of DHCPv6, new deployments are based primarily on DHCPv6.

4.2.3.3 DHCP Server Function   The two PPP tunneling approaches (with or without L2TP) are IP protocol version–independent. While departing from the IPv4 approach, there is no way to assign an IPv6 address or advertise a DNS server address through IPV6CP. A DHCPv6 server function must be fulfilled somewhere in the network. The most obvious approach is to have the endpoint terminating the PPP sessions offering the DHCP server functionality as well. In this way, the subscriber profile contained in the RADIUS attribute can feed the local DHCPv6 server. Another possibility is to make the PPP termination point a DHCPv6 relay agent. The DHCPv6 server function is then fulfilled by a device somewhere else in the network. The advantage of this approach is to leverage the flexibility of a fully featured DHCPv6 server and to avoid the coupling of two protocols (DHCP and RADIUS).

4.2.4 DSL Network Access Technologies

DSL architectures offer a number of alternatives in terms of protocol stacks and are able to accommodate many SP constraints. The commonly deployed access methods are point-to-point protocol over Ethernet (PPPoE), point-to-point protocol over ATM (PPPoA), and routed bridged encapsulation (RBE). The PPPoE, PPPoA, and RBE access methods are good candidates to offer IPv6 connectivity in a variety of deployment models. Methods that are PPP based can leverage the IPv6 extensions to the AAA function and fit well in current IPv4 deployment models.

4.2.4.1 PPPoA Access Model PPP over ATM adaptation layer 5 (AAL5) (RFC 2364) uses AAL5 as the framed protocol, which supports both PVC and SVC. PPPoA was implemented primarily as part of DSL. It relies on RFC1483, operating in either logical link control-subnetwork access protocol (LLC-SNAP) or VC-mux mode. A GWR device encapsulates the PPP session based on RFC1483 for transport across the ADSL loop and the digital subscriber line access multiplexer (DSLAM). Figure 4.5 depicts the global architecture for PPPoA.

4.2.4.2 PPPoE Access Model PPPoE, defined in RFC2516, differs from PPP, as it starts with an endpoint discovery stage. Second, a client–server relation is established between endpoints, whereas PPP is peer to peer in nature. When the endpoint discovery is completed, both ends establish a PPP session. Figure 4.6 depicts the global architecture for PPPoE. The PPPoE session

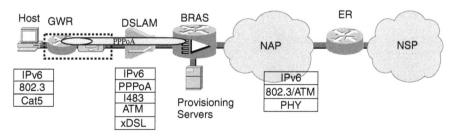

FIGURE 4.5   Architecture for PPPoA access.

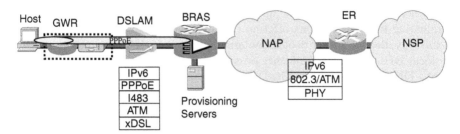

FIGURE 4.6 Architecture for PPPoE access.

originates from the host. When the PPPoE session reaches the GWR, it is encapsulated over ATM. This PPPoEoA traffic is terminated at the BRAS. Since the BRAS is IPv6 enabled, it can natively forward the IPv6 traffic to the ER.

4.2.4.3 RBE Access Model The ATM routed bridged encapsulation routes IP over bridged RFC1483 Ethernet traffic from a bridged LAN (Figure 4.7). The IPv6 packets are bridged to the BRAS and then routed based on the IPv6 header. IPv6 packets are identified based on the Ethertype $0 \times 86dd$. In the case of IPv6, each PVC from the GWR to the BRAS is assigned a different IPv6 prefix. This option offers the simplicity of an Ethernet-based approach. There are no session states retained on the BRAS, resulting in fewer resources needed on the BRAS. The neighbor discovery (ND) cache is maintained on a per interface basis.

4.2.5 PPP Options

For years, the IPv6 introduction in a network has been seen as an additional cost without a clear return on investment (ROI). Since early adopters knew they were deploying IPv6 without short-term ROI, they tried to minimize the cost associated with introducing a new protocol. PPP is particularly well suited for a cost-effective introduction in an SP network. As PPP creates a Layer2

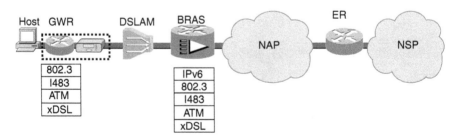

FIGURE 4.7 Architecture for the RBE access model.

connection between the subscriber and the ER, it minimizes the number of nodes that must be IPv6 capable. Using this approach, only the GWR and the ER have to be upgraded. Not all Layer2 devices in the path, such as switches and DSLAM, need to be upgraded. The standard defining IPv6 support of RADIUS attributes was issued by IETF in 2001, making this architecture the path of smooth IPv6 deployment in broadband networks.

4.2.6 Addressing and Routing

A number of addressing options are available in this PPP-based architecture. The main criterion is the size of the address allocation assigned to the subscriber. It varies from a single IPv6 address to a large prefix, well beyond what can be expected from an IPv4 allocation. It is likely that the size of respective IPv4 and IPv6 allocations will be different, leading to different provisioning models for the two protocols, creating additional operational complexities for the SP.

4.2.6.1 PPP-Based Models Routing and addressing considerations are similar for all types of PPP-based models. IPv6 operations will take place wherever the PPP sessions are terminated. In the L2TP-based model described in Figure 4.8, the BRAS does not perform any Layer3 operations for subscriber traffic. In the PPP-terminated model described in Figure 4.9, the BRAS must be fully IPv6-aware in order to forward IPv6 subscriber traffic at Layer3.

4.2.6.2 Allocating a Single IPv6 Address per Subscriber There are several methods of assigning a single IPv6 address per subscriber. The first method is to use the PPP interface identifier to carry the last 64 bits of the IPv6 address. The first 64 bits, the network prefix, are common for a number of subscribers attached to the same BRAS. The same RA is sent on all, or a group, of attached PPP links. By concatenating the PPP interface identifier and the IPv6 network prefix received in the RA, the host creates a 128-bit IPv6 address. In this scenario, all communication between hosts sharing the same prefix has to go through the BRAS, including ND traffic.

The second method of assigning a single IPv6 address per subscriber is to use SLAAC on the host by using the RA sent by the BRAS. In this scenario, all

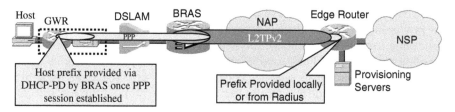

FIGURE 4.8 Addressing scheme in the wholesale deployment model.

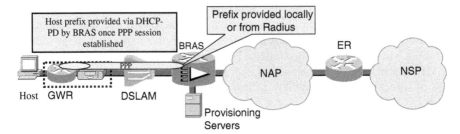

FIGURE 4.9   Addressing scheme in the ISP-operated model.

subscribers share the same IPv6 network prefix. A third way to assign a single IPv6 address to the subscribers is to run DHCPv6. After establishing the PPP session, the subscriber initiates a DHCPv6 transaction and receives a 128-bit IPv6 address. The RA message sent by the BRAS includes the address of the default gateway to be used by the host.

4.2.6.3   Allocating a /64 Prefix per Subscriber   To take full advantage of the large IPv6 address space, an entire /64 prefix can be assigned to each subscriber. A different RA is sent on each PPP link carrying a unique IPv6 prefix. Hosts can use the IPv6 prefix included in the RA message to construct a 128-bit IPv6 address using SLAAC. In a bridged environment, multiple hosts can configure themselves with an IPv6 address using the same IPv6 network prefix.

4.2.6.4   Allocating Multiple Prefixes to Subscribers   In case of a routed GWR, a single /64 prefix allocation is used for assigning addresses to the GWR uplink interface (connected to the BRAS). DHCP-PD offers another shorter prefix (< /64) to the GWR. This prefix is then used by the GWR to assign IPv6 addresses to the downstream interfaces (connected to the hosts). The GWR also sends out RA messages including a /64 prefix on all its downstream interfaces.

4.2.7   Routing Considerations

As mentioned in earlier sections, IPv6 subscribers will typically be assigned a /64 or shorter prefix than will IPv4 subscribers (who are typically assigned a single IP address). The larger IPv6 prefix allocations may require more memory and processing power on the SP network devices. It may also make it more challenging for the SP to manage the size of its routing domain.

Following the guidelines established with IPv4, typically there are no dynamic routing protocols enabled between the subscriber and the first SP Layer3 hop. The subscriber typically uses a default route pointing toward the PPP termination point (BRAS or ER), and vice versa. The default route is installed based on the information included in the RA message sent on the link.

If the same /64 prefix is shared by multiple subscribers, the prefix will be advertised as off-link in the RA message, and the subscribers will forward all the traffic for its local prefix to the SP router using the default route. The GWR will install a default route from the information received in the RA and at the same time perform dynamic routing between upstream (toward the SP Layer3 next hop) and downstream interfaces.

From the SP gateway point of view, the goal is to minimize the number of entries in the routing table while offering optimal routing for subscriber traffic. With IPv6 it is expected that address allocations will typically be static compared to IPv4, where addresses are typically reused. This is applicable as long as the subscriber does not move from its initial location in the network. The SP can allocate a large IPv6 prefix per aggregation point and inject it statically in the routing domain. Each individual subscriber prefix does not need to be manually injected into the routing domain, thus optimizing the size of the SP routing domain.

Due to a physical layer change in the access part of the network, a subscriber circuit may be terminated on a different BRAS. In this case, the ISP-operated deployment model (Figure 4.10) would force subscriber renumbering to keep the routes aggregated. In the case of the wholesale deployment model (Figure 4.11), when the subscriber traffic is always tunneled to the same ER, a physical layer change in the access part of the network does not force renumbering but poses the challenge of a single point of failure (if there is no redundancy on the ER).

When the GWR is a trusted device, there is the possibility of running a simple routing protocol such as RIPng between the GWR and the PPP session termination point (BRAS or ER). Each PPP session will run a different instance of RIPng, and the redistribution of RIPng routes is strictly policed.

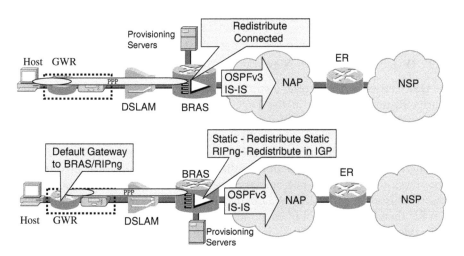

FIGURE 4.10 Routing scheme in the ISP-operated deployment model.

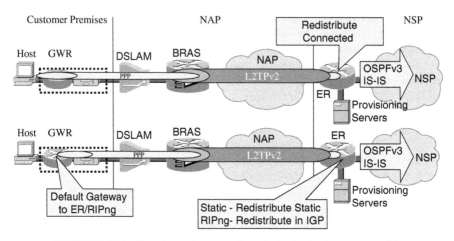

FIGURE 4.11   Routing scheme in the wholesale deployment model.

4.2.8   Routed Bridged Encapsulation

An IPv6 RBE deployment differs from its IPv4 counterpart. For IPv6, a different prefix is assigned to each ATM PVC. Each IPv6 prefix configured on individual PVC is statically assigned by the SP. The GWR acts as a bridge, and hosts are on the same link as the SP router. The SP router sends a RA message corresponding to the prefix on each PVC. SLAAC is used by the hosts to configure themselves with an IPv6 address. If the subscriber connects a GWR router behind the bridged device, DHCP-PD can also be used for prefix allocation. Assignment of individual addresses to hosts is also possible if the BRAS interface running RBE acts as a DHCPv6 server. Regardless of the DHCPv6 functionality used on the BRAS, it will be responsible for injecting aggregated routes in the SP routing domain.

4.2.9   IPv6 Security in DSL Networks

The DSL networks do not offer a specific built-in security scheme similar to the security features available in the DOCSIS cable networks. Vulnerability or inherent protection against attacks is based on the combined Layer2/Layer3 architecture. In all scenarios where PPP is deployed, the network behind the PPP session is identified in the authentication phase and is partially trusted. Although a single IPv6 address or an entire IPv6 prefix can be delegated to the subscriber, the SP does not manage the individual IPv6 addresses. The subscriber is allowed to deploy as many hosts as desired behind its gateway (GWR); there is no control of each individual host.

In this scenario the SP is in charge of applying best practices that block subscribers from using the PPP access to generate attacks, described in RFC2827. Appropriate measures must be taken to block source addresses that are not part of the prefixes allocated. This is called a unicast reverse path forwarding (uRPF) check: For each incoming packet in the SP network, the source address is checked against the routing table. If the route for the source address is not the incoming interface, the packet is dropped. In this way, reflection attacks directed toward the owner of the spoofed source address are defeated.

The DHCPv6 protocol has been complemented with message authentication mechanisms defined in RFC3118. These mechanisms require the distribution of a shared key to the DHCP client, which may not be feasible in all scenarios, especially when the subscriber connects a client that is not managed by the SP. To offer security, DHCP relies on packet filter mechanisms to ensure that the DHCP client–server exchange is secured. For PPP, generally no relay is involved, which simplifies the security model. The DHCP message exchange occurs over a point-to-point link, so there are few possibilities of tampering with the packet contents. Since the PPP link is established using a username/password couple, the DHCP client at the other end of the PPP session is trusted. The PPP session terminator will act as any IPv6 router and implement best practices for dropping all incoming packets, which include a type 0 source routing header as a clear threat to the network.

Ethernet-based access networks are extremely appealing to SPs deploying new infrastructures. Their architecture is simpler and more cost-effective than DSL. A single Layer2 technology is used end to end, which greatly simplifies management and operational costs. IPv6 naturally fits in this design and also for greenfield operators, a dual-stack deployment with parity in design options for both protocols is desirable. DSLAMs, which have been associated with the ATM world, are now migrating to Ethernet uplinks. Based on configuration, the subscriber traffic is encapsulated over one or multiple VLANs. This presents the advantage of keeping the ATM encapsulation only between the subscriber and the DSLAM, simplifying the configuration of BRAS.

4.3   ETHERNET NETWORKS

Today, SPs are faced with the challenge of addressing the needs of their customers by providing innovative and flexible service solutions. Providers that offer residential and small-business services must build and deliver cost-effective, high-quality solutions that exceed today's ability to deliver high-quality voice, video, and data to end users. In certain environments, SPs can offer high-quality services to broadband subscribers using an RJ-45 or fiber to the home (FTTH) 10/100/1000-Mbps Ethernet infrastructure. Such services are generally available in metropolitan areas in multitenant buildings, where an Ethernet infrastructure can be deployed in a cost-effective manner. Since Ethernet is a ubiquitously deployed technology, it provides SP with the ability to deploy IPv6 over a highly reliable and resilient infrastructure.

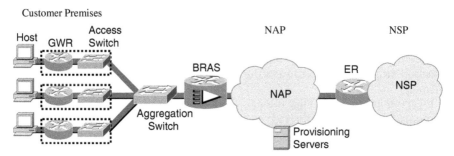

FIGURE 4.12 Ethernet broadband access topology.

4.3.1 Ethernet Network Elements

A typical Ethernet deployment is depicted in Figure 4.12. In scenarios where IPv6 is natively transmitted over Ethernet, there is no absolute need for switches to have IPv6 capability. However, in reality, where the Layer2 network is a key element of the security policy, there is a need for those switches to support the required IPv6 functionality to be part of this security scheme.

The topology is composed of the following elements:

- *Broadband remote access server:* aggregates the user traffic, whether it is IP or PPP encapsulated. It can either handle it at Layer3 or hand it over (L2TP tunnel) to the NSP, except in the wholesale deployment model, where it must be IPv6-capable.
- *Aggregation switch:* aggregates multiple access switches into one high-speed interface configured for 802.1q trunking; maintains Layer2 isolation of customers through VLANs. IPv6 awareness is necessary if it is part of the security scheme, protecting IPv6 address configuration mechanisms.
- *Access switch:* provides access to the aggregation switches in the NAP infrastructure; an optional element.
- *Gateway router:* customer premises gateway router that connects the hosts to the access switches.
- *Host:* PC, notebook, appliance, or other device connected to the GWR.
- *Edge router:* represents the interface between the NAP and the NSP and can be jointly managed; terminates PPP sessions and L2TP tunnels. In all cases, it supports IPv6.

4.3.2 Ethernet Deployment Options

Deployments of Ethernet network elements are different from the DSL architecture as they adopt more flexible and IP-aware Layer2 technologies. Effectively, a number of IPv6 features, such as SLAAC and ND, are designed to run on top of Ethernet. Although it is still possible to stack other access

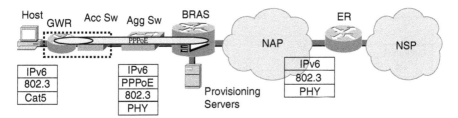

FIGURE 4.13   PPP deployments over an Ethernet infrastructure.

technologies by running PPP over Ethernet, Ethernet infrastructures make IP-based deployments natural and easy.

4.3.2.1   PPP Options   When migrating from an ATM to an Ethernet access infrastructure, it is possible to keep the classical PPP model, to allow for a smooth transition from the DSLAM-based infrastructure to the aggregation switch model. It also allows the SP to keep the same provisioning model regardless of the physical infrastructure. In terms of addressing and routing, there is no difference between a DSL-based model, where PPPoA or PPPoEoA is deployed, and an Ethernet-based model, where PPPoE is used. Figure 4.13 illustrates the Ethernet-based PPP deployment model.

4.3.2.2   Dedicated VLAN Model   The dedicated VLAN model relies on VLANs to isolate each subscriber. Each VLAN is tied to an interface on the BRAS and the GWR. This model (Figure 4.14) offers native access to the full spectrum of IPv6 features.

Addressing and Routing   The addressing and routing model looks similar to the model described in the PPP model. Without PPP, the only way to specify the interface identifier of the GWR uplink interface is to offer a unique IPv6 address through DHCPv6. When DHCP-PD is enabled for the subscriber

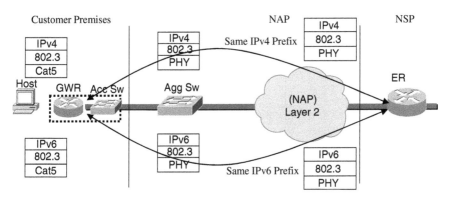

FIGURE 4.14   Per-subscriber VLAN architecture.

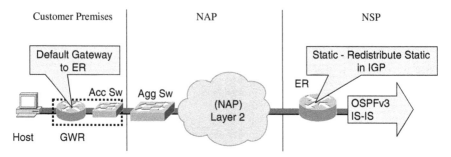

FIGURE 4.15   Routing in a per-subscriber VLAN model.

GWR, the ER is responsible for injecting routes for the prefixes delegated in the SP routing domain. The scenario where the ER acts as a DHCPv6 server is not typical, as the subscriber prefix associations would have to be stored locally on the ER. This is obviously not the preferred solution when thousands of subscribers are attached to the ER. Acting as a DHCPv6 relay agent, the ER will inspect the contents of DHCPv6-relayed packets and will insert and remove matching routes based on the validity of associated leases. These routes will need to be redistributed in the IGP, OSPFv3, or ISIS at the ER, as illustrated in Figure 4.15.

When subscriber-to-prefix binding is stored on a server outside the network, there is a need to identify each subscriber to assign the prefixes. If the subscriber GWR is a trusted device, the server can rely on the DHCPv6 unique identifier (DUID) of the GWR to assign unique IPv6 prefixes if the DUID does not change often. If the subscriber GWR is not a trusted device, the subscriber identification must come from the network. In the dedicated VLAN model, this is relatively easy if the DHCPv6 relay inserts a remote identifier (RFC4649) or subscriber identifier (RFC4580) option that will be used as a key to retrieve the matching prefix (Figure 4.16).

Where the ER acts as a DHCPv6 server, it is likely that the SP will allocate prefixes out of local pools configured on the ER instead of offering fixed IPv6 address allocations. If static subscriber-to-prefix bindings are stored locally, maintenance of this database can be challenging. The most practical solution is

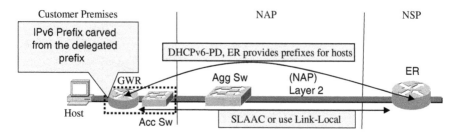

FIGURE 4.16   Addressing in a per-subscriber VLAN model.

to group the configuration of a subscriber-facing interface with its associated delegated prefix.

4.3.2.3   Shared VLAN Model   The shared VLAN model is appealing to SP, as the same VLAN encompasses multiple subscribers. The VLAN tagging standard (802.1q) gives only 12 bytes to identify the VLAN number, offering a total of 4096 VLANs. This can be a problem for large SP routers that aggregate tens of thousands of subscribers. With a shared VLAN model, a 12-byte space is no longer an issue, since multiple subscribers can be associated with a single VLAN. DHCPv6 and ND are not affected particularly by this shared VLAN architecture, as they are designed to work on links populated with a large number of hosts. In the shared VLAN model, the SP would need to upgrade the aggregation switch to Layer3 to implement IPv6 security features and policies to handle customer traffic at Layer3.

4.3.3   Subscriber Identification

In the Ethernet model, subscriber identification authentication offered by PPP may not be available unless the PPPoE model is used. There are other mechanisms available, such as port-level security via 802.1X authentication and MAC-based filtering, which can be used in other deployment models. As physical access is of little value without a valid IP address, DHCP uses different mechanisms to identify the subscriber. The SP network device, which is connected to the subscriber, intercepts the client's DHCP message and adds DHCP options that help to identify the subscriber's physical location. This mechanism offers a rudimentary form of identification and is a good compromise that has been widely deployed.

A similar model can be used for IPv6; the SP network device connected to the subscriber can intercept the DHCPv6 packet and add such identification options as remote identifier (RFC4649) or subscriber identifier (RFC4580). In the current state of the standard, there is no provision in DHCPv6 for insertion of options by nodes that are not acting as Layer3 devices. The only place where DHCPv6 options can be inserted is by the DHCPv6 relay agent, which is acting as a Layer3 device. The DHCPv6 relay must source the relayed packet with an IPv6 address.

One possible way to identify subscribers at Layer2 using DHCP would be to define a new set of DHCPv6 options that can be inserted by Layer2 devices connected to the subscriber. These DHCPv6 options can be used for topological identification of the subscriber and need to be excluded from any message hashing used for DHCP authentication as defined in RFC3118.

4.3.4   IPv6 Security in Ethernet Networks

In IPv6, the use of SLAAC for address assignment makes it difficult to manage and track hosts. Autoconfiguration of host addresses also existed in IPv4 but was not an integral part of the protocol design. As we have seen, the Ethernet LAN can be dedicated to a subscriber or shared among multiple subscribers. Those two models present different security issues. Where an Ethernet LAN is dedicated to a single subscriber, the SP needs to protect itself against malicious subscribers that would use their access to attack the SP infrastructure or the Internet. In this case, the SP acting as a good Internet citizen protects others from malicious users. Dedicated access to the SP infrastructure makes the subscriber a trusted device in the network. In the ISP-operated model there is very little opportunity for interception of attacks, as all the devices between the subscriber's home and the SP network are owned by the same entity.

Where an Ethernet LAN is shared among multiple subscribers, SPs are adopting IPv6 policies that do not differ much from their IPv4 counterpart. The subscriber-to-subscriber traffic is forwarded to the Layer3 next hop, where security policies are enforced and Layer2 ↔ Layer3 address resolution takes place. No traffic is forwarded directly from one subscriber to the other. The ER must ensure that ND functions such as DAD work on such a NBMA link. This can be achieved by implementing ND proxy (RFC4389) or inverse ND (RFC3122) mechanisms on the ER. Regardless of the VLAN model chosen, the SP will deploy security measures at the Layer3 device connected to the subscribers to protect the subscribers and its own infrastructure.

The IPv6 ND protocol is very dynamic, which makes it an ideal target for attackers. The SP will need to make sure that ND mechanisms are properly protected to handle scanning attacks on the attached VLANs. Scanning itself must be prevented, but above all, the protective measures against scanning must not turn into a DoS attack where the Layer3 node is overloaded by performing excessive address resolution.

In the Ethernet deployment model, the aggregation switch presents a first barrier of protection against spoofing attacks; however, it is a good practice, especially when not all switches are IPv6-aware, to deploy security mechanisms such as uRPF on the Layer3 device connected to the subscribers.

4.4   IEEE 802.11A/B/G WIRELESS NETWORKS

A number of IPv6 features, such as SLAAC and the inherent large address allocations, make it a good candidate for wireless networks. As an example, a single /64 prefix can be used for provisioning an entire LAN without exhausting the address space. Two different models are available for IPv6 deployment in wireless networks. In the first model the NAP operates at Layer2 only and all IPv6 traffic is forwarded to the NSP. In the second model, the NAP infrastructure is Layer3-aware and IPv6 routing is enabled in the NAP. The NAP also takes part in the delegation of IPv6 prefixes from the NSP.

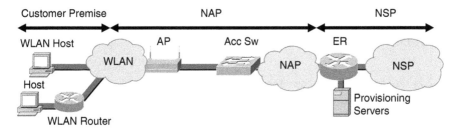

FIGURE 4.17   Layer2 network access provider model.

4.4.1   Wireless Network Elements

In wireless networks, the access point is a Layer2 device providing network access to subscribers. However, in order to deploy IPv6 in wireless networks a number of devices (from the subscriber to the NSP) need to be IPv6 aware. The following elements constituting the wireless access architecture are illustrated in Figure 4.17.

- *Access point:* a Layer2 device that provides connectivity between a WLAN host/router and the NAP. The AP should be able to process Layer2 frames with a frame protocol ID of $0 \times 86DD$, indicating IPv6.
- *Access switch:* a Layer2 device that connects the AP to the ER. It can be owned by the NAP or the NSP.
- *Access router:* a router that connects the AP to the ER; can be owned by the NAP or the NSP.
- *Edge router:* an upstream router to the AR/access switch that connects to the NSP backbone.
- *WLAN host:* PC, notebook, PDA, or other device with an IPv6-enabled wireless network interface card (NIC).
- *WLAN router:* router with a IPv6-enabled wireless NIC that connects a host to the WLAN cloud.

4.4.2   Layer2 NAP with Layer3 Termination at ER

When offering wireless service in this model, the NAP operates at Layer2 only and all IPv6 traffic is forwarded to the NSP. This deployment model is illustrated in Figure 4.16. IPv6 services can easily be deployed in this model, as the NAP infrastructure remains IPv6 unaware. One must verify that forwarding of Ethernet frames with IPv4 Ethertype (0×800) is not hard-coded in any of the network devices, which prevents the forwarding of other Ethertype, such as IPv6 ($0 \times 86dd$). It is likely that direct subscriber-to-subscriber communication will not be allowed in a public deployment to enforce policies at Layer3 on the ER. Conversely, the access switch must implement similar security policies for IPv4 and IPv6. Security features need to be implemented to protect IPv6 mechanisms such as SLAAC and DHCPv6 against DoS attacks and other activities.

4.4.2.1 Addressing In the absence of PPP, the wireless subscriber can establish a direct Layer2 connection with the network, making the wireless deployment model very similar to the wired deployment model (Figure 4.18). Two solutions exist for subscriber IPv6 address provisioning. The first option is to use SLAAC for IPv6 address assignment and stateless DHCPv6 to obtain other network parameters, as the RA message does not offer the flexibility of carrying several DHCPv6 options. This option is well suited for purposely uncontrolled environments such as large conventions, conferences, and so on, due to its simplicity and economical benefits. The second option is to use stateful DHCPv6 for IPv6 address assignment as well as for obtaining other network parameters. DHCPv6 can also be used as a central tracking tool and can participate well in the network audit scheme. This functionality is particularly needed in a public wireless environment, where a connection can be short-lived and can be originated from a loosely identified source.

The ER on the link sends RA messages with the appropriate settings to control the wireless hosts' provisioning mode. The RA message can include information about the link-local address of the default router, which can be used by the hosts for installing a default route to the next hop. A single /64 prefix can be shared across all wireless hosts that are part of the same VLAN. The RA message can also include the prefix length of the IPv6 prefix to be used by the host.

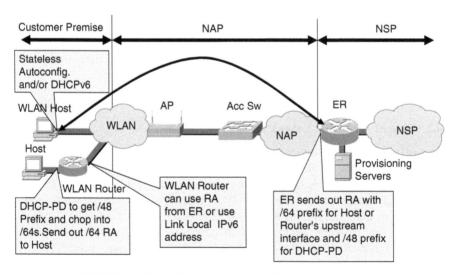

FIGURE 4.18   Addressing in a Layer2 NAP environment.

To ease configuration of the WLAN router, one can leverage DHCP-PD. The DHCP-PD prefix can be delegated by the DHCPv6 server or the ER acting as a DHCP-PD delegating router. The shorter-than /64 prefix is divided in /64 chunks to number the WLAN router's downstream interfaces automatically. On each downstream interface, SLAAC can be used for IPv6 address configuration by all hosts that are connected.

The ER can be configured manually with /64 prefixes on its downstream interfaces connected to the WLAN routers. The ER sends out RA messages, including the IPv6 prefixes, to be used by the WLAN routers for configuring their upstream interfaces with an IPv6 address using SLAAC. The WLAN routers can also use DHCPv6 for acquiring an IPv6 address for their upstream interfaces.

These addressing schemes can ease customer network renumbering, as all the WLAN router interfaces are configured automatically. Provided that IPv6 customer renumbering is easier, the wireless SP can relax its constraints on maintaining static address allocations to subscribers.

4.4.2.2 Routing The general rule for public access networks is to avoid routing adjacencies between subscriber routers and the SP routing infrastructure. To mitigate this, the subscriber equipment installs a default route toward the SP infrastructure from the RA it receives from the ER (Figure 4.19). The ER that performs DHCPv6 relay functionality or acts as a delegating router for DHCP-PD inserts static routes corresponding to prefixes delegated to subscribers. These static routes, individually or aggregated, have to be injected in the SP routing domain to direct return traffic to the ER.

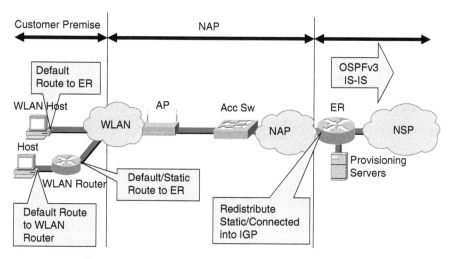

FIGURE 4.19 Routing in a Layer2 NAP environment.

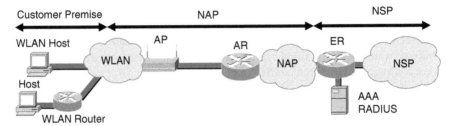

FIGURE 4.20   Layer3 access provider model.

4.4.3   Layer3-Aware NAP with Layer3 Termination at AR

While wireless design based on a Layer2 access switch (AS) forces the use of VLAN to multiplex the traffic coming from subscribers, design based on a Layer3 access router (AR) offers more IP-based options (Figure 4.20). For PPP-based models, PPP sessions are initiated from the WLAN host and terminated on the AR, or initiated from the AR and terminated an the ER. IP-based solutions can rely on IPv6 VPN (RFC4659) to direct customer traffic pertaining to a particular NSP from the NAP access points to the edge router. The NAP acts as a VPN provider for its NSP clients. The traffic from each NSP is carried in a particular VPN, switched over the MPLS network through the NAP instrastructure, and is finally delivered to the edge router. MPLS-based solutions are covered in more detail in Section 6.1.2.3.

4.4.3.1 PPP-Based Model   PPP-based models that are being deployed widely for other access technologies can also be used for wireless networks (Figure 4.21).

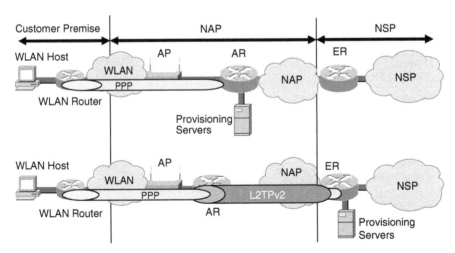

FIGURE 4.21   PPP-based model in a wireless environment.

The SP can leverage provisioning systems used in other parts of the network and deploy IPv6 in a very cost-effective manner. The use of different provisioning models per access technology can be an expensive proposition for the SP. The ISP-operated model or the wholesale model can be deployed in an IPv6 WLAN environment using PPPoE technologies. To support the ISP-operated model, the AR receives the IPv6 traffic at Layer3 and needs to be upgraded to support IPv6. To support the wholesale deployment model, the AR does not need to support IPv6, as it tunnels the PPP session to the ER using L2TPv2. To support IPv6 the ER needs to be upgraded in both models.

4.4.3.2 Addressing and Routing The PPP models create tunnels that are easily replicable over any physical topology. Overall, PPP routing and addressing schemes are similar for all access technologies. In the case of wireless technologies, which can be used in a number of mobility scenarios, connecting the subscriber networks from various places over time can result in route deaggregation. This may become an issue for the SP in the long run. As the subscriber moves from one hot spot to another with its fixed IPv6 address allocation, the SP infrastructure must deploy routing or tunneling mechanisms to direct the subscriber traffic back to its current location. The addressing and routing considerations for the ISP-operated and wholesale deployment models in the wireless environment are very similar to the DSL environment. These addressing and routing schemes have been discussed in detail in Sections 4.2.6 and 4.2.7, respectively. Therefore, these details are not covered again in this section.

4.4.4 IPv6 Security in Wireless Networks

When PPP is used for deploying IPv6 over a wireless infrastructure, all security recommendations related to IPv6 still apply. For the most part, the wireless deployment model is similar to the Ethernet model except that the transient and mobile nature of the wireless subscriber connections makes attackers difficult to identify. The wireless infrastructure must enforce security for the IPv6 provisioning mechanisms. The ND and DHCPv6 messages must be monitored and protected. For example, no subscriber should be allowed to act as a DHCPv6 server. If autoconfigured addresses are allowed as source addresses, the Layer2 infrastructure must prevent the theft of IPv6 addresses among subscribers.

4.5 SUMMARY

Commercial IPv6 products appeared in the market around 2001. The original products were focused primarily on the core of the network, where deployment options were limited. Later enhancements of routing protocols to support IPv6 provided a necessary boost to future deployments. SPs implemented these

IPv6-capable routing protocols, and operational networks appeared. Shortly after that, the IETF community worked on standards for enabling IPv6 on PPP-based networks. Since then, this architecture has been deployed by many SPs worldwide with remarkable success. With PPP-based networks currently representing a significant share of all broadband networks in a number of countries, the path to widespread deployment of IPv6 is well paved.

SPs are considering adopting Ethernet-based networks, migrating away from PPP, thus placing many expectations on DHCPv6 and Layer2 switches. As a control protocol with no hardware dependencies, DHCPv6 is evolving rapidly to address challenges in "PPP-less" environments. Layer2 switches are (in shared network architecture) partly responsible for enforcing network security. To operate at wire speed, switches need to process packets in hardware. In most cases, their role in IPv6 security will require new hardware to support IPv6. This will probably place short-term restrictions on architectural options for Ethernet-based IPv6 broadband networks.

Standardization efforts in the broadband IPv6 area are still in progress. As more players deploy IPv6, they come up with specific needs for their network environments, which are then included in the standardization process.

REFERENCES

1. Y. Shirasaki, S. Miyakawa, T. Yamasaki, and A. Takenouchi, "A Model of IPv6/IPv4 Dual Stack Internet Access Service," RFC4241, December 2005.
2. B. Parker, "The PPP AppleTalk Control Protocol (ATCP)," RFC1378, November 1992.
3. W. Simpson, "The PPP Internetwork Packet Exchange Control Protocol (IPXCP)," RFC1552, December 1993.
4. W. Simpson, "The Point-to-Point Protocol (PPP)," RFC1661, July 1994.
5. D. Haskin and E. Allen, "IP Version 6 over PPP," RFC2472, December 1998.
6. G. McGregor, "The PPP Internet Protocol Control Protocol (IPCP)," RFC1332, May 1992.
7. J. Solomon and S. Glass, "Mobile-IPv4 Configuration Option for PPP IPCP," RFC2290, February 1998.
8. G. Gross, M. Kaycee, A. Lin, A. Malis, and J. Stephens, "PPP Over AAL5," RFC2364, July 1998.
9. J. Heinanen, "Multiprotocol Encapsulation over ATM Adaptation Layer 5," RFC1483, July 1993.
10. L. Mamakos, K. Lidl, J. Evarts, D. Carrel, D. Simone, and R. Wheeler, "A Method for Transmitting PPP Over Ethernet (PPPoE)," RFC2516, February 1999.
11. P. Ferguson and D. Senie, "Network Ingress Filtering: Defeating Denial of Service Attacks Which Employ IP Source Address Spoofing," RFC2827, May 2000.
12. R. Droms and W. Arbaugh, "Authentication for DHCP Messages," RFC3118, June 2001.

13. B. Volz, "Dynamic Host Configuration Protocol for IPv6 (DHCPv6) Relay Agent Remote-ID Option," RFC4649, August 2006.

14. B. Volz, "Dynamic Host Configuration Protocol for IPv6 (DHCPv6), Relay Agent Subscriber-ID Option," RFC4580, June 2006.

15. D. Thaler, M. Talwar, and C. Patel, "Neighbor Discovery Proxies (ND Proxy)," RFC4389, April 2006.

16. A. Conta, "Extensions to IPv6 Neighbor Discovery for Inverse Discovery," RFC3122, June 2001.

17. J. De Clercq, D. Ooms, M. Carugi and F. Le Faucheur, "BGP-MPLS IP Virtual Private Network (VPN) Extension for IPv6 VPN," RFC4659, September 2006.

18. S. Asadullah and A. Ahmed, "IPv6 in Broadband," Cisco Systems, Inc. Packet Magazine, Fourth Quarter 2004.

19. S. Asadullah, A. Ahmed, C. Popoviciu, P. Savola, and J. Palet, "ISP IPv6 Deployment Scenarios in Broadband Access Networks," RFC4779, January 2007.

20. B. Lourdelet, "Application Note: IPv6 Access Services," Cisco Systems, Inc.

21. B. Lourdelet, "Application Note: DHCPv6," Cisco Systems, Inc.

22. S. Asadullah and A. Ahmed, "BRKIPM-3300: Service Provider IPv6 Deployment," Cisco Networkers 2007, January 2007.

23. A. Ahmed, C. Popoviciu, and J. Palet, "IPv6 Deployment Scenarios in Broadband Access Networks," Global IPv6 Summit, Spain, June 2005.

5 Configuring and Troubleshooting IPv6 on Gateway Routers and Hosts

In this chapter we discuss configuring and troubleshooting IPv6 on gateway routers (GWRs) and hosts that reside in customer premises or home networks. We also provide an overview of how IPv6 can be enabled on Windows XP/Vista, Windows Server 2003 and 2008, Linux, MAC OS, and Sun Solaris. For IPv6 to be widely deployed and accepted by the user community it is important for various operating systems not only to support IPv6 provisioning and configuration, but also to support IPv6-based applications. The SP needs to provide IPv6-based value-added services, such as high-quality data and voice and video over IP, to end customers to get the user community excited about IPv6 and adopt its use.

One of the key advantages of IPv6 to end customers is that they can get a /64 dedicated prefix assigned to them which does not have to change unless they change their provider. With IPv4, users typically get a single public IPv4 address assigned to them by the SP via DHCPv4. The IPv4 address typically has a DHCP lease time of a few hours or days. If the user switches off its PC for an extended period of time, so that the IPv4 address lease is not renewed with the DHCP server, the address is returned to the available pool of addresses, which can be assigned to a different user. If end customers want to host an application or a server and need a fixed IPv4 address, they typically pay extra for this service, due to the associated administrative overhead. Additional charges also typically apply when end customers want more than a single IPv4 address for their homes or businesses.

Since IPv6 has an enormous amount of address space available, the SP can assign a dedicated /64 prefix to end customers who have a single network in their homes. For customers who have multiple networks in their homes or for business customers, the SP can allocate shorter than the /64 prefix via DHCP-PD. Since there is no shortage of IPv6 prefixes today, these prefixes can typically be assigned permanently to these customers. (*Note*: The permanence of the prefix has an impact on SP routing table deaggregation in case of SP

Deploying IPv6 in Broadband Access Networks, By Adeel Ahmed and Salman Asadullah
Copyright © 2009 John Wiley & Sons, Inc.

infrastructure changes or partial subscriber mobility. Therefore, user prefix allocation persistence is not a given in all cases.)

Another advantage of IPv6 includes enabling true peer-to-peer applications without the use of network address translation (NAT), which breaks transparency and makes management of end devices difficult for the SP. Today, NAT is deployed widely, due to the limited number of IPv4 addresses available. End customers typically deploy a GWR between their hosts and the device (DSL modem, cable modem, Ethernet switch, etc.) that connects them to the SP network. Deploying a GWR allows the customer to receive a single IP address from the SP and connect multiple devices behind it. The GWR receives a public (routable) IP address on its interface connected to the SP device/network and is configured with a private IP address on the interface connected to the customer's home network. The GWR can act as a lightweight DHCP server for devices in the home network and assign them private IP addresses from a configured range. It then provides address translation from the private IP addresses to the public IP addresses and vice versa for devices trying to connect to the SP network or the Internet. This deployment model is illustrated in Figure 5.1.

NAT is also perceived as a security feature by end customers since they do not need to expose their actual IPv4 address to the outside world and can hide their network behind a GWR or a NAT device. With IPv6 there should not be a need to use NAT to connect to the outside world. This may become a security issue for some users, as they may no longer be unable to hide their network behind a NAT device or GWR. IPv6 has several ways to address these security concerns. One method is to use IPv6 privacy extensions as described in RFC3041. Other methods can include enabling firewall capability on hosts and GWR, and the methods outlined in RFC4864. Let's take a look at how IPv6 can be enabled on GWR and hosts, and the various IPv6 deployment options available on different operating systems.

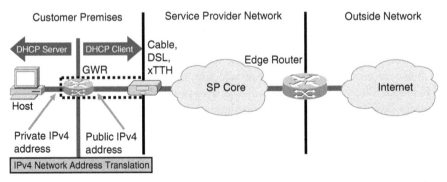

FIGURE 5.1 Gateway router deployed on customer premises to provide IPv4 NAT.

5.1   IPv6 SUPPORT ON GATEWAY ROUTERS

If the SP does not offer IPv6 services to its customers, several options are available to end users to obtain IPv6 connectivity. Customers can begin by upgrading hosts and other devices to dual-stack mode. By upgrading the hosts to dual-stack status, customers can connect to other IPv6 networks, including the IPv6 Internet, by tunneling their IPv6 traffic over the existing IPv4 SP infrastructure. A dual–stack GWR deployed between their hosts and the device that connects them to the SP network can provide customers with multiple options for tunneling IPv6 traffic from the GWR to other IPv6 devices.

Figure 5.2 illustrates a typical IPv6 deployment scenario where the customer has deployed a dual-stack GWR to connect to the IPv6 Internet over an IPv4 infrastructure.

In the example above, the end customer has deployed a dual-stack GWR to connect to the IPv6 Internet by tunneling all IPv6 traffic over the SP IPv4 infrastructure. The GWR receives a public IPv4 address from the SP DHCP server on its WAN interface connected to the SP network. It acts as a DHCP client on its upstream (WAN) interface and acts as a DHCP server on its downstream (LAN) interface, as illustrated in Figure 5.1. The dual-stack host on the LAN receives its private IPv4 address from the GWR and uses the GWR as its default gateway for forwarding all IPv4 traffic.

The end customer can sign up for any of the IPv6 services available on the Internet to get IPv6 connectivity. Based on the type of IPv6 service the customer receives, the GWR can be configured accordingly to initiate a tunnel to the IPv6 service router to carry IPv6 traffic over the SP network. The tunnel is terminated on the IPv6 service router, which can route the customer's IPv6 traffic to the IPv6 Internet, and vice versa. One thing to note is that both tunnel

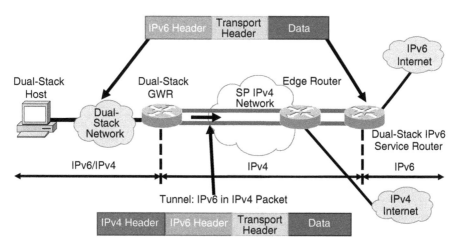

FIGURE 5.2   Connecting to the IPv6 Internet by deploying a dual-stack GWR.

endpoints (source and destination) can be the IPv4 addresses of the GWR WAN interface and the dual-stack IPv6 service router. Once the tunnel comes up on IPv4, the GWR can receive an IPv6 address on its tunnel interface via either SLAAC or DHCPv6. For manual tunnels, the IPv6 address can also be statically assigned to the GWR. If the customer has multiple LAN segments connected to the GWR, the customer can request shorter than /64 prefixes from the IPv6 service provider. The GWR can act as a DHCP-PD requesting router (RR) and request a prefix to be delegated from the IPv6 service provider. The prefix can be assigned from a DHCP-PD delegating router (DR) acting as a DHCPv6 server or a DHCPv6 server located elsewhere in the SP network. Once the GWR receives the delegated prefix, it can subdivide the prefix into smaller chunks of /64 and send out RA messages on its LAN interfaces advertising the /64 IPv6 prefixes that can be used by the hosts to configure themselves with an IPv6 address. The hosts can use the GWR as their IPv6 default gateway. The GWR can forward traffic between the host and the IPv6 Service Router. (*Note*: The protocol number for IPv6 in IPv4 tunneled traffic is 41. If the SP does not specifically block protocol-41 in its network, customers can tunnel their IPv6 traffic over the SP infrastructure even without the SP's knowledge.)

If the SP is able to offer native IPv6 services to its customers, tunneling is not required. To connect to other IPv6 networks customers can upgrade their hosts and GWR to dual-stack mode. To communicate with other IPv6 devices the hosts and GWR will need to be provisioned with IPv6 addresses. The provisioning options for hosts are discussed in detail in Section 5.2. The provisioning options for GWR are discussed in this section. Figure 5.3 illustrates the native IPv6 deployment model for end customers using a dual-stack GWR.

In this model, all network components are typically dual-stack, including the hosts, GWR, edge router, and provisioning and management servers. The SP core could be dual-stack or IPv4-only. The mechanisms for transporting IPv6 traffic over the SP core network are described in detail in Chapter 6.

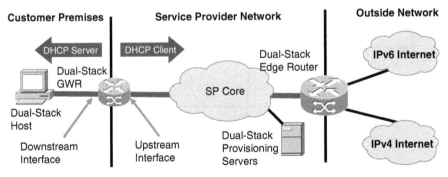

FIGURE 5.3 Native IPv6 deployment model for end customers.

As illustrated in Figure 5.3, the GWR and host on the customer's premises are upgraded to dual-stack mode to deploy IPv6 in the customer's network. The GWR acts as a DHCP client on its upstream interface connected to the SP network and as a DHCP server on its downstream interface connected to the customer network. The GWR still receives an IPv4 address on its upstream interface connected to the SP network. This address is typically a public (routable) address. On its downstream interface, the GWR is configured with a private IPv4 address and is also configured with a DHCP pool for handing out private IPv4 addresses to hosts connected to it.

For the IPv6 address, the GWR can either use its link-local IPv6 address or can receive a global unicast or unique-local IPv6 address on its upstream interface. The global unicast or unique-local IPv6 address is assigned via either SLAAC or DHCPv6. (The GWR typically looks for RA messages on its upstream interface, sent by the next-hop IPv6 router, to determine how to assign itself an IPv6 address.) This address is in addition to the IPv6 link-local address, which is assigned to the GWR upstream interface automatically when IPv6 is enabled. The GWR can use the IPv6 link-local, or global unicast, or unique-local address to communicate with its upstream router. The GWR also acts as a DHCP-PD RR and requests a prefix to be delegated from a DHCP-PD DR or a DCHPv6 server. This delegated prefix is typically shorter than /64. Once the GWR receives the prefix via DHCP-PD, it can carve it into smaller chunks of /64 and assign it to its downstream interfaces. The GWR sends out RA messages with an A-bit and an O-bit set to 1 on its downstream interfaces advertising these /64 prefixes. The hosts connected to the GWR downstream interfaces receive the /64 prefix via the RA messages and configure themselves with an IPv6 address using SLAAC. The hosts then use stateless DHCPv6 to acquire other network parameters, such as the address of the DNS server. The GWR can either reply back to the hosts with the configuration information requested (e.g., the address of the DNS server) or it can also act as a DHCPv6 relay and forward the request to the upstream DHCPv6 server.

Since numerous GWR implementations are on the market, it is difficult to cover every GWR configuration in this book. Therefore, the authors have chosen to use a Cisco IOS-based GWR configuration example for illustration purposes. The behavior recommended for GWR is documented in an IETF draft [17]. Following is an example of DHCP-PD client configuration using a Cisco IOS-based GWR.

```
ipv6 unicast-routing
! Enable IPv6 routing in global configuration mode.
ipv6 dhcp pool IPv6
! Set up the local DHCP pool for IPv6.
dns-server 2001:db8:dead:beef:a00:20ff:fee5:6403
! IPv6 DNS server address.
domain-name v6.foobar.com
! IPv6 domain name.
```

```
!
interface
 ipv6 address v6Prefix 0:0:0:1::/64 eui-64
```
! The first 48 bits of the IPv6 address come from DHCP-PD, so they are listed as 0:0:0 here. The first subnet, which is :1, is assigned to this interface.
```
 ipv6 nd other-config-flag
```
! Set the O-bit to 1 in the RA message.
```
 ipv6 nd ra-interval 30
```
! Send out RA messages every 30 seconds.
```
 ipv6 dhcp server IPv6
```
! Act as a DHCPv6 server on this interface.
```
!
interface
 ipv6 address autoconfig default
```
! Use SLAAC for configuring an IPv6 address.
```
 ipv6 nd ra suppress
```
! Do not send out RA messages on this interface.
```
 ipv6 dhcp client pd v6Prefix
```
! Act as a DHCP-PD RR on this interface.
```
 ipv6 rip RIP enable
```
! Enable IPv6 RIP or RIPng on this interface. This enables RIPng between the GWR upstream interface and the next-hop IPv6 router.
```
!
ipv6 router rip RIP
```
! Configure the RIPng process in global configuration mode.
```
 redistribute connected
```
! Redistribute the connected prefixes to the upstream IPv6 router.

For the upstream router configuration and complete DHCPv6 debugs, refer to Chapter 6.

The reason for enabling RIPng (RFC2080) between the GWR and the next-hop IPv6 router is to advertise the delegated prefixes to the upstream router. Since the upstream router may only act as a DHCP-PD relay, it may not be aware of the prefixes that are delegated to the GWR. This is one way of advertising the delegated prefixes to the upstream router to fix that issue. Running an IPv6 IGP between the GWR and the next-hop IPv6 router is typically common in cable networks, as discussed in Chapter 3.

Another way for the upstream router to learn about the delegated prefixes is to manually snoop the DHCP-PD messages between the GWR and the DHCP-PD DR or DHCPv6 server and install a static route based on the information gleaned from the snoop for the delegated prefixes in its routing table pointing to the GWR's upstream interface.

As mentioned previously, the type of IPv6 deployment options available on GWR may vary from one vendor to another, so many options are available to end customers looking at deploying IPv6 in their networks. IPv6 support on hosts also varies from one operating system to another. In the next few sections we take a look at IPv6 support on different OSs and the deployment options available on each OS.

5.2 IPv6 SUPPORT ON WINDOWS XP, WINDOWS VISTA, AND WINDOWS SERVER 2003 AND 2008

Microsoft Windows XP Service Packs 1 and 2 (SP1 and SP2), Windows Vista, and Windows Server 2003 and 2008 can all be provisioned and managed using IPv6. To connect them to the IPv6 Internet or other IPv6 networks, these operating systems support various IPv6 features that can be used for configuration. There are differences in behavior and configuration requirements among the various OS versions. For example, IPv6 needs to be enabled manually on Microsoft Windows XP SP1 and SP2 and Windows Server 2003, whereas Microsoft Windows Vista and Windows Server 2008 have IPv6 enabled by default. In the examples that follow, no manual configuration would be required to enable IPv6 and various features on the later platforms. The following commands can be used from the command prompt to enable IPv6 manually:

```
C:\>ipv6 install
OR
C:\>netsh interface ipv6 install
```

To uninstall IPv6, the following commands can be used:

```
C:\>ipv6 uninstall
OR
C:\>netsh interface ipv6 uninstall
```

It is expected that IPv6-capable hosts will have more than one IPv6 address assigned to one or more of their interfaces at any given time. Once IPv6 is enabled on a Windows host, it can configure itself automatically with a link-local IPv6 address by using the EUI-64 format as defined in Section 2.3. Once the IPv6 host configures its link-local IPv6 address, it sends out an RS message to the all-routers multicast address (FF02::2) to discover any routers present on the local link. If a router is present on the local link, it responds back with an RA message and includes the global unicast or unique-local IPv6 prefix to be used by the host to configure itself with an IPv6 address. It also includes other options, which instruct the host whether to use SLAAC or DHCPv6 to configure an additional IPv6 address, and how to request other network parameters, such as a DNS server address. The RA message can be a unicast back to the requesting host or can be a multicast message to the all-nodes multicast address (FF02::1) so that other hosts present on the link can also be informed about the presence of an IPv6-enabled router. The router needs to be configured with the correct settings, on the interface connected to the host, to send out RA messages with the appropriate flags.

At this time, Windows XP SP1 and SP2 and the Windows 2003 server do not support stateful DHCPv6 and can use SLAAC only for IPv6 address assignment and stateless DHCPv6 to acquire other network parameters. Windows Vista and the Windows 2008 server do support stateful DHCPv6 and can request an IPv6 address as well as other network parameters using this provisioning method. Since Windows XP is more widely deployed than Windows Vista at this time, it poses an issue for SPs that would like to use stateful DHCPv6 to provision end devices connected to their network. In this case, end customers running Windows XP on their hosts would need to install an application such as the Dibbler (see ref. 5), which runs on top of the Windows XP stack and supports stateful DHCPv6.

5.2.1 IPv6 Deployment Options on Windows XP, Windows Vista, and Windows Server 2003 and 2008

Several options are available for deploying IPv6 on the Windows OS. If the SP has the ability to provide native IPv6 connectivity to its customers, no special configuration would be needed on hosts running Windows OS to provide IPv6 connectivity. The Windows host would be provisioned using the techniques discussed in Section 5.2 and can be managed using IPv6 transport.

If the SP does not provide IPv6 connectivity to its customers, the end users have several options available on Windows hosts to connect to the IPv6 Internet. The customer can use any of the tunneling mechanisms supported in Windows OS, such as the Intra-Site Automatic Tunnel Address Protocol (ISATAP) (RFC5214), Teredo (RFC4380), 6to4 tunneling (RFC3056), IPv6 tunnel broker (RFC3053). to tunnel IPv6 traffic over the SP's IPv4 infrastructure and connect to the IPv6 Internet.

What is interesting is that the SP may not even be aware of this kind of traffic, since all IPv6 packets can be encapsulated in IPv4 and transported over the SP infrastructure just like other IPv4 packets. This may be another reason for the SP to provide IPv6 services to its end customers rather than having end customers tunnel IPv6 traffic over its IPv4 infrastructure. The SP can choose to block this IPv6 traffic at Layer2 using the IPv6 frame protocol ID ($0 \times 86DD$) or at Layer3 using IP protocol 41.

It is important to understand the behavior of dual-stack hosts when IPv6 is enabled. For example, Windows Vista and Server 2008 prefer IPv6 over IPv4 and try to initiate communications using IPv6. Upon initialization they first assign themselves a link-local IPv6 address and then they try to discover any IPv6-enabled routers on the link by sending out a RS message. If a router is present on the link, it can respond back with a RA that contains information for the hosts on how to configure themselves with a global unicast or unique-local IPv6 address by using either SLAAC or DHCPv6. If no router is present or the host is not able to configure a global unicast or unique-local IPv6 address, the host then tries to connect to the IPv6 network using ISATAP or 6to4 tunneling. If no ISATAP- or 6to4-enabled routers are present in the network, the host then tries to use Teredo as a last resort to connect to the IPv6 network. If Teredo service is unavailable, the host falls back to IPv4 and uses its IPv4 address to establish communication over the IPv4 infrastructure.

If the host is provisioned in dual-stack mode, it can initiate communication using IPv6 and/or IPv4. In the end the decision to use IPv6 or IPv4 is up to the application. If the host performs a DNS query and receives a response with both an A record (IPv4 address) and an AAAA record (IPv6 address) in the reply, it is up to the application to decide which address to use to initiate communication. This behavior may vary between different OSs as well as different applications.

For example, Windows Vista performs both A and AAAA queries to determine the best method of connectivity to the desired endpoint. The DNS client service in Windows Vista has been designed to perform DNS queries as follows (DNS Behavior in Vista); see ref. 22):

- If the host has only link-local or Teredo IPv6 addresses assigned, the DNS client service sends a single query for A records only.
- If the host has at least one IPv6 address assigned that is not a link-local or Teredo address, the DNS client service sends a DNS query for A records and then a separate DNS query to the same DNS server for AAAA records. If an A record query times-out or has an error (other than the name not found), the corresponding AAAA record query is not sent.

In the following sections we discuss how various tunneling techniques apply to Windows hosts. Although several options are available on Windows hosts for IPv6 deployment, only a few are mentioned in this chapter and discussed in detail.

5.2.1.1 6to4 Tunneling on Windows Hosts   6to4 is an automatic tunneling technique used to connect multiple isolated IPv6 sites over an IPv4 infrastructure. The well-known global unicast IPv6 prefix for 6to4 tunneling is 2002::/16, and the 6to4 address format is the 2002: < 32-bit IPv4 address converted into hexadecimal format > ::/48. The 32-bit IPv4 address in the 6to4 address is taken from the 6to4 router's outgoing interface facing the IPv4 network, as shown in Figure 5.4. The 2002: < 32-bit IPv4 address in hexadecimal format > ::/48 prefix is divided into multiple /64 prefixes and assigned to the 6to4 router's interfaces connected to the IPv6 network. The 6to4 router sends out RA messages on the interfaces configured with the /64 prefixes so that hosts connected to these interfaces can use the /64 prefixes and configure themselves with a 6to4 IPv6 address. Hosts use SLAAC to configure themselves with an IPv6 address.

Devices within the same 6to4 network can communicate with each other directly using their 6to4 IPv6 address. If a device within one 6to4 network wants to communicate with a device in another 6to4 network, it must initiate a tunnel from the local 6to4 router, over the IPv4 infrastructure, to a 6to4 router at the destination site. Once the 6to4 tunnel comes up, all IPv6 traffic is transported over this tunnel by encapsulating IPv6 packets in IPv4. The tunnel destination address is embedded within the 6to4 IPv6 destination address and the tunnel is created dynamically; therefore, only a single tunnel interface is required on the local 6to4 router for initiating tunnels to multiple destinations. This is a very scalable approach for deploying IPv6 when multiple isolated IPv6 sites need to be connected over the IPv4 infrastructure. Only the hosts and edge routers need to be upgraded to support IPv6; all other intermediate devices remain IPv4-only.

Figure 5.4 shows a 6to4 tunnel connecting multiple IPv6 sites over an IPv4 infrastructure. In the figure, the 6to4 tunnel is established over the IPv4 Internet, so the 6to4 routers need to have a public IPv4 address assigned to the interface connected to the IPv4 Internet. All IPv6 traffic is encapsulated in IPv4 and is tunneled over the IPv4 infrastructure. The IPv4 destination address of the 6to4 tunnel is embedded in the IPv6 packets sent by the hosts. For example, if host A wants to communicate with host C, it will send IPv6 traffic with the IPv6 source address of 2002:d1a8:6301:1::1 and a destination IPv6 address of 2002:d1a8:1e01:3::3. When this traffic reaches the local 6to4 router, it looks at the 6to4 IPv6 destination address and extracts from it the

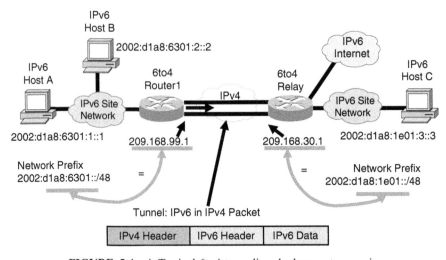

FIGURE 5.4 A Typical 6to4 tunneling deployment scenario.

IPv4 destination address. Since d1a8:1e01 is equal to 209.168.30.1 in dotted decimal format, the local 6to4 router knows the tunnel destination address by converting the 32 bits (bits 17 to 48) in the 6to4 IPv6 destination address to dotted decimal format. Once it knows the tunnel destination address it can establish a 6to4 tunnel using a source IPv4 address of 209.168.99.1. This is its own IPv4 address of the outgoing interface and it uses a destination IPv4 address of 209.168.30.1, which is the address it extracted from the IPv6 destination address. This is how the 6to4 router can dynamically set up tunnels to multiple destinations using the same local tunnel interface and the same source IPv4 address.

In another example, if host A wants to connect to the IPv6 Internet and initiates a 6to4 tunnel, it just needs to know the 6to4 relay router's IPv6 address, so it can forward the non-6to4 traffic to this IPv6 address. All IPv6 traffic (6to4 as well as other IPv6 traffic) goes over the same 6to4 tunnel interface. At the remote end, the 6to4 relay router looks at the destination IPv6 address and forwards the IPv6 traffic accordingly. If the IPv6 destination address is a 6to4 address, it forwards the traffic to the IPv6 network connected. If the IPv6 destination address is a non-6to4 address, it forwards the IPv6 traffic to the IPv6 Internet. Using this tunneling technique, users can connect to other 6to4 networks as well as the IPv6 Internet as long as a 6to4 relay service is available.

Enabling 6to4 Tunneling on Windows Hosts   On Windows hosts, 6to4 tunneling can be enabled using CLI commands from the command prompt. The **netsh** commands shown in Figure 5.5 can be used to enable 6to4 service on Windows hosts.

The commands shown in Figure 5.6 can be used to verify 6to4 configuration on Windows hosts. To verify if the host received a 6to4 address from the router, you can look at the output of **ipconfig** from the command prompt (Figure 5.7).

5.2.1.2 ISATAP Tunneling on Windows Hosts
ISATAP is a tunneling technique used primarily in enterprise networks to provide IPv6 connectivity to hosts. The host just needs to know the IPv4 address of the ISATAP router that can provide it with an IPv6 prefix. The host can then configure itself with an IPv6 address. The IPv4 address for the ISATAP router can be configured manually on the host or the network administrator can create an A record for "ISATAP" on the DNS server. Once the IPv6-enabled host boots up, it assigns itself a link-local address and tries to verify the presence of an IPv6-enabled router on the link by sending out an RS message. If the host does not receive an RA and does not have a global unicast or unique-local address, it tries to use ISATAP tunneling to connect to the IPv6 network. The host first tries to locate the IPv4 address of the ISATAP router to which it can connect. Once it knows the address of the ISATAP router, it initiates an ISATAP tunnel with the ISATAP router using its IPv4 address. Once the ISATAP tunnel comes up, the host can request an IPv6

```
C:\>netsh interface ipv6 6to4 set ?
The following commands are available:

Commands in this context:
set interface    - Sets 6to4 interface configuration information.
set relay        - Sets 6to4 relay information.
set routing      - Sets 6to4 routing information.
set state        - Sets the 6to4 state.

C:\>netsh interface ipv6 6to4 set interface "Local Area Connection" enabled
Ok.
! Enables 6to4 service on the specified interface.

C:\>netsh interface ipv6 6to4 set state state=enabled undoonstop=disabled
Ok.
! Specifies whether 6to4 state is enabled and if 6to4 is disabled on service stop.

C:\>netsh interface ipv6 6to4 set routing routing=enabled
Ok.
! Specifies whether 6to4 routing is enabled.

C:\>netsh interface ipv6 6to4 set relay 6to4.ipv6.org. enabled
Ok.
! Specifies the name of the 6to4 relay service. This would require a resource record
entry in the DNS service so the 6to4 host can map this to an IPv6 address.
```

FIGURE 5.5   6to4 Configurations on Windows hosts.

prefix from the ISATAP router so that it can configure itself with an IPv6 address. Figure 5.8 illustrates an ISATAP deployment model.

The ISATAP address format is [< 64-bit Network Prefix > :0000:5efe: < 32-bit IPv4 address >]. The 64-bit network prefix is sent by the ISATAP router to

```
C:\>netsh interface ipv6 6to4 dump
! This command displays the 6to4 configuration on the host
# ----------------------------------
# 6to4 Configuration
# ----------------------------------
pushd interface ipv6 6to4

reset
set state state=enabled undoonstop=disabled
! Set 6to4 state to enabled on the host
set relay name=6to4.ipv6.org. state=enabled interval=1440
! 6to4 relay information
set routing routing=enabled
! Enabling 6to4 routing on the host
set interface name="Local Area Connection" routing=enabled
! Define the interface to be used for 6to4 tunneling

popd
# End of 6to4 configuration
```

FIGURE 5.6   Verifying 6to4 configuration on Windows hosts.

```
C:\>ipconfig
:
Tunnel adapter 6to4 Tunneling Pseudo-Interface:

    Connection-specific DNS Suffix  . : ipv6.test.com
    IP Address. . . . . . . . . . . : 2002: d1a8:6301:1:215:58ff:fe7e:9a06
    Default Gateway . . . . . . . . : 2002:d1a8:6301:1:200:cff:fe3a:8b18
```

FIGURE 5.7   Verifying 6to4 address assignment on windows hosts.

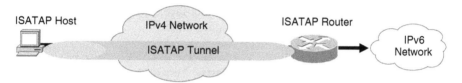

FIGURE 5.8   ISATAP deployment model.

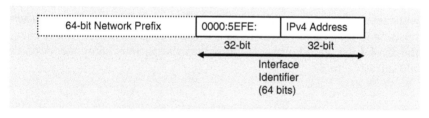

FIGURE 5.9   ISATAP address format.

```
C:\>netsh interface ipv6 isatap set router 192.168.10.1
```

FIGURE 5.10   ISATAP configuration on Windows XP/Vista host.

the host over the ISATAP tunnel. The 32-bit 0000:5efe is well known OUI for ISATAP, and the last 32-bit address is the IPv4 address of the host's interface used to establish the ISATAP tunnel. Figure 5.9 shows the ISATAP address format.

ISATAP Configuration on Windows Hosts   ISATAP tunneling can be enabled on Windows hosts using **netsh** from the command prompt. Figure 5.10 shows how ISATAP can be configured manually on a Windows XP/Vista host.

```
C:\>ipconfig
:
Tunnel adapter Automatic Tunneling Pseudo-Interface:
Connection-specific DNS Suffix  . :
IP Address. . . . . . . . . . . . : 2001:db8:f00d:1:2:5efe:192.168.30.10
IP Address. . . . . . . . . . . . : fe80::5efe:192.168.30.10%2
Default Gateway . . . . . . . . . : fe80::5efe:192.168.10.1%2
```

FIGURE 5.11   Verifying ISATAP operation on Windows XP/Vista host.

To verify that the host has received an IPv6 prefix from the ISATAP router and has configured an IPv6 address on its interface, you can use the **ipconfig** command from the command prompt (Figure 5.11).

5.2.1.3 Teredo Support on Windows Hosts   Teredo is an IPv6 transition technology that provides address assignment and host-to-host automatic tunneling for unicast IPv6 traffic when hosts are located behind one or multiple IPv4 network address translators (NATs). To traverse IPv4 NATs, IPv6 packets are sent as IPv4-based UDP messages. IPv6 traffic tunneled using Teredo can cross one or multiple NATs and allow a Teredo client to access the hosts on the IPv6 Internet (through a Teredo relay) and other Teredo clients on the IPv4 Internet. Figure 5.12 illustrates the various components of the Teredo infrastructure.

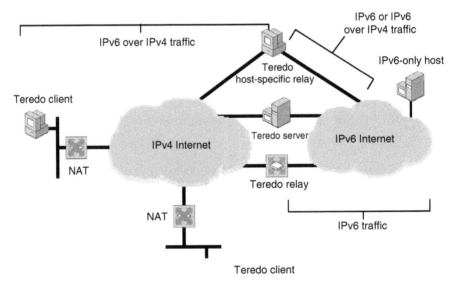

FIGURE 5.12   Components of the Teredo infrastructure.

Teredo Client   A Teredo client is a dual-stack node that supports a Teredo tunneling interface through which packets are tunneled to other Teredo clients or nodes on the IPv6 Internet (via a Teredo relay). A Teredo client communicates with a Teredo server to obtain an address prefix from which a Teredo-based IPv6 address is configured or to help initiate communication with other Teredo clients or hosts on the IPv6 Internet.

Teredo Server   A Teredo server is a dual-stack node that is connected to both the IPv4 Internet and the IPv6 Internet and supports a Teredo tunneling interface over which packets are received. The primary role of the Teredo server is to assist in the address configuration of a Teredo client and to facilitate the initial communication between Teredo clients and other Teredo clients or between Teredo clients and IPv6-only hosts. The Teredo server listens for Teredo traffic on UDP port 3544.

Teredo Relay   A Teredo relay is a dual-stack router that can forward packets between Teredo clients on the IPv4 Internet (using a Teredo tunneling interface) and IPv6-only hosts. In some cases, the Teredo relay interacts with a Teredo server to help it facilitate initial communication between Teredo clients and IPv6-only hosts. The Teredo relay listens for Teredo traffic on UDP port 3544.

Teredo Host-Specific Relay   A Teredo host-specific relay is a dual-stack node that has an interface and connectivity to both the IPv4 Internet and the IPv6 Internet and can communicate directly with Teredo clients over the IPv4 Internet, without the need for an intermediate Teredo relay. Connectivity to the IPv4 Internet can be through a public IPv4 address or through a private IPv4 address and using NAT. The connectivity to the IPv6 Internet can be through a direct connection to the IPv6 Internet or through an IPv6 transition technology such as 6to4, where IPv6 packets are tunneled across the IPv4 Internet. The Teredo host-specific relay listens for Teredo traffic on UDP port 3544.

Teredo Address Format   The Teredo address format is illustrated in Figure 5.13. The first 32 bits are the well-known prefix for Teredo 2001::/32. The next 32 bits are the global IPv4 address of the Teredo server that helped configure the Teredo address for the client. The next 16 bits are reserved for Teredo flags, which help define the type of NAT the client is behind. For Windows Vista– and Windows

32 bits	32 bits	16 bits	16 bits	32 bits
Teredo Prefix	**Teredo Server IPv4 Address**	**Flags**	**Obfuscated External Port**	**Obfuscated External Address**

FIGURE 5.13   Teredo address format.

Server 2008–based Teredo clients, unused bits within the flags field provide a level of protection from address scans by malicious users. The next 16 bits store an obscured version of the external UDP port corresponding to all Teredo traffic for this Teredo client. The last 32 bits store an obscured version of the external IPv4 address corresponding to all Teredo traffic for this Teredo client. This is the public IPv4 address used by NAT. Refer to RFC4380 for more details on Toredo addressing and operation.

Teredo Configuration on Windows Hosts Teredo can be enabled on a Windows host using the **netsh** configuration from the command prompt, as illustrated in Figure 5.14. To verify Teredo, the command illustrated in Figure 5.15 can be used. To verify if the Teredo client received an IPv6 address, the command illustrated in Figure 5.16 can be used.

If the host is not able to receive an IPv6 address or the Toredo server is unreachable over UDP, check to see if the connection is being blocked by a firewall. If running Windows Vista, an IPv6 firewall is required for Toredo to operate.

```
C:\>netsh interface ipv6 set teredo client
```

FIGURE 5.14   Enabling a Teredo client on windows hosts.

```
C:\>netsh interface ipv6 show teredo
Teredo Parameters
-----------------------------------------------
Type                          : client
Server Name                   : teredo.ipv6.test.com
Client Refresh Interval       : default
Client Port                   : default
State                         : probe(cone)
Type                          : teredo client
Network                       : unmanaged
NAT                           : cone
```

FIGURE 5.15   Verifying a Teredo configuration on windows hosts.

```
C:\>netsh interface ipv6 show address
Interface 5: Teredo Tunneling Pseudo-Interface

Addr Type  DAD State   Valid Life   Pref. Life   Address
---------  ----------  -----------  -----------  ----------------------------------------
Public     Preferred   infinite     infinite     2001:0:4136:e37e:0:fbaa:b97e:fe4e
Link       Preferred   infinite     infinite     fe80::ffff:ffff:fffd
```

FIGURE 5.16   Verifying a Teredo address configuration on Windows hosts.

5.3   IPv6 SUPPORT ON LINUX

There are two main IPv6 implementations for Linux: the implementation that comes as part of the Linux kernel and the UniverSAl playGround for IPv6 (USAGI) implementation. The USAGI project works to deliver a production-quality IPv6 protocol stack for Linux, tightly collaborating with the KAME [18], WIDE [19], and TAHI [20] projects. Some of the Linux OSs that currently support IPv6 include Red Hat Linux 6.2 and higher, Mandrake 8.0 and higher, SuSE 7.1 and higher, and Debian 2.2 and higher. In this section we look at how IPv6 can be enabled on Linux OSs and how Linux hosts can be configured for IPv6 deployment. (*Note*: The examples and illustrations used in this section are based on Red Hat Linux. Other implementations may have a different way of enabling and configuring IPv6.)

5.3.1   Deploying IPv6 on Linux

To deploy IPv6 on Linux, we first need to enable IPv6 on the host. Once IPv6 is enabled, the host can configure itself with an IPv6 address using SLAAC or DHCPv6. If native IPv6 connectivity is available, the host can connect directly to the IPv6 network using its global unicast or unique-local IPv6 address, and no further configuration may be required on the host. If there is no native IPv6 connection to the IPv6 network or the IPv6 Internet, the host's IPv6 traffic will need to be tunneled over the IPv4 infrastructure to connect to the IPv6 network.

5.3.1.1   Enabling IPv6 on Linux   Let's take a look at how IPv6 can be enabled on a Linux host. To enable IPv6 the user needs to edit the /etc/sysconfig/network file, add the NETWORKING_IPV6 = yes entry, and then restart networking or reboot the host as illustrated in Figure 5.17.

Once IPv6 is enabled, the host should be able to configure itself with an IPv6 address. The **ifconfig ⟨interface name⟩** command can be used to verify the IPv6 address on the host. This is illustrated in Figure 5.18.

In this example we can see that the host configured itself with the link-local IPv6 address fe80::240:f4ff:fe6c:c8af using the EUI-64 format and its hardware address 00:40:f4:6c:c8:af. The host received a 2001:db8:dead:beef::/64 IPv6 prefix and configured itself with the global unicast IPv6 address 2001:db8:dead:beef:240:f4ff:fe6c:c8af. The host should be able to use its global unicast IPv6 address to communicate with other devices in the IPv6 network. If

```
# vi /etc/sysconfig/network

NETWORKING_IPV6=yes

! Save and reboot
```

FIGURE 5.17   Enabling IPv6 on Linux.

```
# ifconfig eth0
eth0    Link encap:Ethernet  HWaddr 00:40:F4:6C:C8:AF
        inet addr:10.1.1.100  Bcast:10.1.1.255  Mask:255.255.255.0
        inet6 addr: 2001:db8:dead:beef:240:f4ff:fe6c:c8af/64 Scope:Global
        inet6 addr: fe80::240:f4ff:fe6c:c8af/10 Scope:Link
        UP BROADCAST RUNNING MULTICAST  MTU:1500  Metric:1
        RX packets:289223 errors:0 dropped:0 overruns:0 frame:0
        TX packets:13452 errors:0 dropped:0 overruns:0 carrier:0
        collisions:0 txqueuelen:100
        RX bytes:53425777 (50.9 Mb)  TX bytes:3381080 (3.2 Mb)
        Interrupt:5 Base address:0xf000
```

FIGURE 5.18   Verifying the IPv6 address on the Linux host.

the host is connected to an IPv4-only network, it needs to be configured with a tunneling technique, which can be used to connect the host to the IPv6 network.

5.3.1.2   *Tunneling IPv6 on Linux*   There are several tunneling mechanisms available on Linux to support IPv6 deployment. General tunneling mechanisms are discussed in this section. The type of tunneling mechanism depends on the type of network the user is connected to and the type of service available. If the user is connected to an enterprise network, ISATAP can be used. If the user is connected to an SP network, configured tunnels may have to be used to connect to the IPv6 network.

Figure 5.19 illustrates how IPv6-configured tunnels can be enabled on Linux hosts. Since this is a manually configured tunnel, the source and destination IPv4 addresses need to be statically defined. The user needs to know the IPv4 address of the device that will terminate this tunnel. Another thing to note is that this is a point-to-point tunnel, so a new tunnel interface would need to be

```
# ip tunnel add sit1 mode sit remote 209.10.1.2 local 190.168.1.100
! Create the tunnel interface. Remote address is the IPv4 address of the router
terminating the   tunnel. Local address is the IPv4 address of the Linux client.

# ip link set sit1 up
! Enable the tunnel interface

# ip address add dev sit1 2001:db8:feed:f00d::2/64
! Assign an IPv6 address to the tunnel interface.

# ip route add ::/0 dev sit1
! Add a default route for all IPv6 traffic to go over the tunnel interface.
```

FIGURE 5.19   Enabling configured tunnels on linux hosts.

```
#ip tunnel show sit1
sit1: ipv6/ip  remote 209.10.1.2  local 190.168.1.100  ttl inherit

#route -A inet6 | grep sit1
Kernel IPv6 routing table
Destination              Next Hop      Flags  Metric  Ref   Use   Iface
2001:dB8:feed:f00d::/64  ::            UA     256     10    0     sit1
fe80::/10                ::            UA     256     6     0     sit1
ff02::9/128              ff02::9       UAC    0       1     0     sit1
ff00::/8                 ::            UA     256     0     0     sit1
::/0                     ::            U      1024    0     0     sit1

# ip -6 addr show sit1
6: sit1@NONE: <POINTOPOINT,NOARP,UP> mtu 1480 qdisc noqueue
    inet6 fe80::a5e:a64d/128 scope link
    inet6 2001:db8:feed:f00d::2/64 scope global
```

FIGURE 5.20   Verifying operation for configured tunnels.

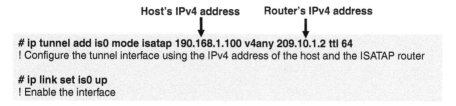

FIGURE 5.21   ISATAP configuration on Linux hosts.

created for every IPv6 site to which this host needs to connect. To verify if the tunnel is operational, the commands illustrated in Figure 5.20 can be used. Another tunneling option available on Linux hosts is ISATAP. Figure 5.21 illustrates how ISATAP can be configured on Linux hosts. The ISATAP tunnel is sourced from 190.168.1.100 and is terminated at the ISATAP router at 209.10.1.2. Once the tunnel comes up, the host can request an IPv6 prefix from the router and configure itself with an IPv6 address. As ISATAP is more geared toward enterprise networks, it may not commonly be deployed by end customers connected to an SP network. Details about ISATAP are discussed in Section 5.2.1.2.

5.4   IPv6 SUPPORT ON MAC OS X

MAC OS X 10.2 and higher has IPv6 installed and active by default. MAC OS X IPv6 implementation is based on the KAME stack. IPv6 can be configured via either the graphical user interface (GUI) network tab or the command line. MAC OS X supports several IPv6 features, including SLAAC using EUI-64 format for an interface identifier, tunneling mechanisms such as configure tunnels and 6to4, IPv6 DNS [MAC OS X prefers IPv6 if a DNS query results in IPv4 (A record) and IPv6 (AAAA record) addresses being returned], and IPv6 privacy extensions. At the time of writing, no DHCPv6 support was available on MAC OS X. Since the MAC OS is based on Unix (BSD), the configuration options are very similar when working with the command line.

5.4.1 Enabling IPv6 on MAC OS X

As mentioned earlier, IPv6 can be enabled on MAC OS X via either a command line or the GUI. Figure 5.22 illustrates how to enable and disable IPv6 on MAC OS X via a command line. MAC OS X 10.2 and higher allows you to enable IPv6 using the GUI from the **System Preferences→Network** control panel. Figure 5.23 illustrates how IPv6 can be enabled via the GUI. The IPv6 address can be configured manually or automatically by using SLAAC. If IPv6 is enabled by default, the MAC host should have a link-local

```
# To enable IPv6 on all interfaces:

ip6 -a

# To disable IPv6:

ip6 -x
```

FIGURE 5.22 Enabling and disabling IPv6 on MAC OS X via a command line.

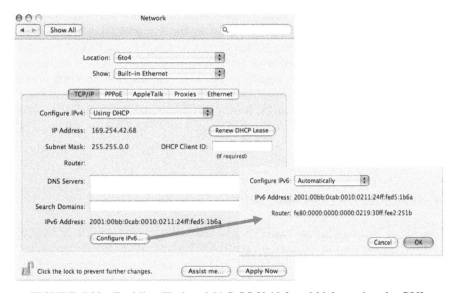

FIGURE 5.23 Enabling IPv6 on MAC OS X 10.2 and higher using the GUI.

address assigned to its interface using EUI-64 format for the interface ID. The global unicast or unique-local can be assigned from the IPv6 prefix advertised in the RA message since at present the DHCPv6 is not supported. Other network parameters, such as the DNS server address, can also be configured manually or can be learned via stateless DHCPv6 if the O-bit is set in the RA message received.

5.4.2   Tunneling IPv6 on MAC OS X

MAC OS X supports configured tunnels and 6to4 tunnels with the generic tunnel interface (gif). Setting up a manual tunnel requires several steps on the command line. Figure 5.24 illustrates how configured tunnels can be enabled using the gif. To verify the configuration in the figure, the **ifconfig** command can be used to look at the tunnel interface IP addresses. Figure 5.25 illustrates the configuration of the gif.

MAC OS X also supports 6to4 tunneling if the MAC host needs to connect to multiple IPv6 destinations such as other 6to4 sites and the IPv6 Internet. 6to4 tunneling can be configured using the GUI under **System Preferences→ Network→Location→Network Port Configurations**. There should be an option for 6to4. If the option does not show up, the 6to4 option can be selected by clicking on the **New...** tab and then the **Port** pull-down menu (Figure 5.26). Now 6to4 should show up as one of the check boxes under **Network Port**

```
# ifconfig gif0 tunnel create
! Create the tunnel interface

# ifconfig gif0 tunnel 190.168.2.101 209.30.1.2
! Setup the IPv4 endpoints of the tunnel. 190.168.2.101 is the local IPv4 address of the MAC
client and 209.30.1.2 is the remote IPv4 address of the router that terminates the tunnel.

# ifconfig gif0 inet6 alias 2001:db8:feed:f00d::2
! Configure IPv6 address of the tunnel interface.

# route add -inet6 default -interface gif0
! Add a default route for IPv6 pointing to the tunnel interface.
```

FIGURE 5.24   Example of configured tunnels on a MAC OS X.

```
# ifconfig gif0

gif0: flags=8051<UP,POINTOPOINT,RUNNING,MULTICAST> mtu 1280
      tunnel inet 190.168.2.101 --> 209.30.1.2
      inet6 fe80::203:93ff:feee:9f1f prefixlen 64 scopeid 0x2
      inet6 2001:db8:feed:f00d::2 prefixlen 64
```

FIGURE 5.25   Configured tunnel example on a MAC OS X.

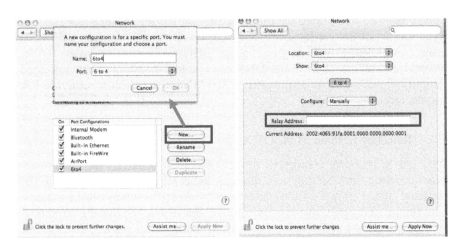

FIGURE 5.26 6to4 configuration on a MAC OS X.

Configurations. Click on the **Edit** tab and change the **Configure** menu from **Automatically** to **Manually** to specify the 6to4 router address.

5.5 PPPv6 SUPPORT ON MAC OS X

MAC OS X integrates the same IPv6-capable PPP daemon as Linux and Solaris. The PPP client is configured and controlled via the GUI network control panel (Figure 5.23) under **PPPoE**. Configure IPv6 has to be set to **Automatically**. When dialing with the PPPoE tool (in the menu bar), the PPP daemon pppd tries automatically to negotiate IPv6 while establishing the PPP link (IPv6CP) and successfully creates a link-local address for the PPP interface. The IPv6 stack afterward receives RA messages on the PPP link and configures a global unicast or unique-local IPv6 address on the PPP interface. After establishing a connection, the default route (for IPv6) has to be put on the PPP interface (ppp0, for example) (Figure 5.27).

5.6 IPv6 SUPPORT ON SOLARIS

IPv6 is supported on Solaris 8, 9, and 10. At the time of installation, the user is prompted for IPv6 activation. If IPv6 is not activated during the installation process, it can be activated manually after installation is complete. Solaris supports several IPv6 features, such as DHCPv6, SLAAC, IPv6 privacy extensions, and DNS for IPv6. Solaris prefers IPv6 if a DNS query results in IPv4 and IPv6 addresses being sent back (A and AAAA RRs).

```
# route add -inet6 default -interface ppp0
! Add a default route pointing to interface ppp0.
```

FIGURE 5.27   Adding a default route for PPP.

```
# touch /etc/hostname6.hme0
! Create a hostname6.<interface name> file to manually activate IPv6 on a specific interface. In
this example the interface name is hme0.

# reboot
! Restart the Solaris host for the changes to take effect. The host should now have a Link-local
IPv6 address.

# ifconfig –a6
! Use this command to verify if the interface received an IPv6 address.
hme0: flags=2000841 mtu 1500 index 2
        ether 8:0:20:56:a8
        inet6 fe80::a00:fe73:56a8/10
```

FIGURE 5.28   Activating an IPv6 on Solaris manually.

5.6.1 Manually Activating IPv6 on Solaris

IPv6 can be manually activated for each interface with the creation of the file /etc/hostname6.⟨*intf*⟩ (*intf* being the name of the interface) (Figure 5.28). For regular IPv6 operation the file can be left empty. However, configuration options can be added for more complex functionality. After creating the file, the system has to be restarted. Upon booting, the network stack sends out RS messages to discover if an IPv6-enabled router is present on the link. If an RA is received, the host tries to configure a global unicast or unique-local IPv6 address using SLAAC and the EUI-64 format for an interface identifier.

5.6.2 Tunneling IPv6 on Solaris

Solaris supports multiple tunneling mechanisms, such as configured tunnels and 6to4 tunnels for connecting IPv6 hosts/devices across an IPv4 infrastructure. Figure 5.29 descries the steps used to create configured tunnels. To verify the configuration, use the **ifconfig –a6** command as shown in Figure 5.30.

The IPv4 source address of the tunnel is 190.168.100.1 and the destination address is 209.30.1.2. Once the tunnel is established, all IPv6 traffic is encapsulated in IPv4 and carried over the IPv4 infrastructure. The IPv6 address of the tunnel interface (tun0:1) is 2001:db8:feed:f00d::2/64 and the destination next-hop address is 2001:db8:feed:f00d::1, which is the tunnel interface on the device that is terminating this tunnel. This is a point-to-point tunnel between the Solaris host and another IPv6-enabled device. If the Solaris host needs to communicate with multiple IPv6 destinations, it needs to be connected to a 6to4 router as discussed in Section 5.2.1.1.

```
# ifconfig ip.tun0 inet6 plumb
! Create the tunnel interface.

# ifconfig ip.tun0 inet6 tsrc 190.168.100.1 tdst 209.30.1.2 up
! Define the source and destination IPv4 address of the tunnel. The source address is the host's
address while the destination address is for the router which will terminate the tunnel.

# ifconfig ip.tun0 inet6 addif 2001:db8:feed:f00d::2/64 2001:db8:feed:f00d::1 up
! Add IPv6 address to the tunnel interface and define the destination next-hop IPv6 address.
Once the tunnel interface comes up all IPv6 traffic will be forwarded over it to the IPv6 next-hop.
```

FIGURE 5.29   Enabling configured tunnels on Solaris.

```
ip.tun0: flags=2200851<UP,POINTOPOINT,RUNNING,MULTICAST,NONUD,IPv6> mtu 1480 index 3
        inet tunnel src 190.168.100.1 tunnel dst  209.30.1.2
        tunnel hop limit 60
        inet6 fe80::4065:406a/10 --> fe80::a5e:a644
ip.tun0:1: flags=2200851<UP,POINTOPOINT,RUNNING,MULTICAST,NONUD,IPv6> mtu 1480 index 3
        inet6 2001:db8:feed:f00d::2/64 --> 2001:db8:dead:beef::1
```

FIGURE 5.30   Verifying configured tunnel operation on Solaris.

```
# touch /etc/hostname6.ip.6to4tun0
! Configure a 6to4 pseudo-interface on the router by creating the /etc/hostname6.ip.6to4tun0 file.

# vi /etc/hostname6.ip.6to4tun0
tsrc 209.30.1.2 2002:d11e:102:1::2/64 up

! Edit the hostname6.ip.6to4tun0 file; add the IPv4 source address of the 6to4 tunnel and the
6to4 IPv6 address of the tunnel interface to it and save it.
! tsrc <IPv4-address> <2002:IPv4-address:subnet-ID:interface-ID:/64> up


# reboot
! Reboot the 6to4 router for the configuration to take effect.
```

FIGURE 5.31   Configuring a 6to4 router on solaris.

A Solaris host can also be configured as a 6to4 router. Figure 5.31 illustrates how to configure a Solaris host as a 6to4 router. After the reboot, a properly configured 6to4 router should have an IPv4 address as well as the 6to4 IPv6 address assigned to the 6to4 pseudointerface as illustrated in Figure 5.32. The 6to4 tunnel is created dynamically when the 6to4 router sees IPv6 traffic with a 2002::/16 destination. The IPv4 destination address of the tunnel is embedded in the IPv6 header of the packet, as discussed in detail in Section 5.2.1.1. If the 6to4 router needs to communicate with a 6to4 relay, it needs to be configured for this operation. Figure 5.33 illustrates how the 6to4 router can be configured to communicate with a 6to4 relay.

```
# ifconfig ip.6to4tun0 inet6

ip.6to4tun0: flags=2200041<UP,RUNNING,NONUD,IPv6>mtu 1480 index 11
        inet tunnel src 209.30.1.2
        tunnel hop limit 60
        inet6 2002:d11e:102:1::2/64
```

FIGURE 5.32   Verifying 6to4 router configuration on Solaris.

```
# /usr/sbin/6to4relay –e
! The -e option sets up a tunnel between the 6to4 router and an anycast 6to4 relay router.
Anycast 6to4 relay routers have the well-known IPv4 address 192.88.99.1. The anycast relay
router that is physically nearest to your site becomes the endpoint for the 6to4 tunnel. This relay
router then handles packet forwarding between your 6to4 site and a native IPv6 site.

# /usr/sbin/6to4relay –e –a <6to4 Relay Router's Address>
! Use this option to specify a different 6to4 relay IP address.

# /usr/sbin/6to4relay
6to4relay: 6to4 Relay Router communication support is enabled.
IPv4 destination address of Relay Router=192.88.99.1
! Use this option to verify 6to4 relay configuration.
```

FIGURE 5.33   Configuration for communicating with a 6to4 relay on Solaris.

5.7   TROUBLESHOOTING IPv6 ON GWR AND HOSTS

Our intent in this section is not to discuss IPv6 troubleshooting in great detail, as it can be a vast topic and is outside the scope of this book. Therefore, we have selected a few examples to illustrate the troubleshooting approach to IPv6-related issues rather than trying to cover every troubleshooting scenario that may exist on GWR and hosts. Several tools are available for trouble-shooting IPv6-related issues on GWR and hosts. The first thing to keep in mind is that while troubleshooting IPv6-related problems one has to apply a generic troubleshooting approach which may be common to both IPv4 and IPv6. For example, if the user is not able to connect to the IPv6 Internet or access an IPv6 website, it is important to understand what could be causing this problem. The issue could be with the user's PC or laptop (MTU settings, etc.). It could be with the web browser or the application being used. It could also be related to IPv6 provisioning or simply just an IPv6 connectivity (routing) issue between the host and the IPv6 Internet.

So the first thing to do while troubleshooting IPv6-related issues is to narrow the scope of the problem. This could be as simple as verifying IPv6 connectivity between the two endpoints either by using an IPv6 ping and/or trace route or by trying to connect to a different endpoint. Figure 5.34 illustrates the output of IPv6 ping and trace route.

```
C:\Documents and Settings\adahmed>ping 2001:420:A00:FFDA::2

Pinging 2001:420:a00:ffda::2 with 32 bytes of data:

Reply from 2001:420:a00:ffda::2: time=202ms
Reply from 2001:420:a00:ffda::2: time=225ms
Reply from 2001:420:a00:ffda::2: time=257ms
Reply from 2001:420:a00:ffda::2: time=182ms

Ping statistics for 2001:420:a00:ffda::2:
    Packets: Sent = 4, Received = 4, Lost = 0 (0% loss),
Approximate round trip times in milli-seconds:
    Minimum = 182ms, Maximum = 257ms, Average = 216ms

C:\Documents and Settings\adahmed>
C:\Documents and Settings\adahmed>
C:\Documents and Settings\adahmed>
C:\Documents and Settings\adahmed>
C:\Documents and Settings\adahmed>tracert 2001:420:A00:FFDA::2

Tracing route to 2001:420:a00:ffda::2 over a maximum of 30 hops

  1    114 ms    133 ms    119 ms  2001:420:1:fff:0:5efe:171.69.7.186
  2    197 ms    151 ms    176 ms  2001:420:a00:ffda::2

Trace complete.
```

FIGURE 5.34   Verifying IPv6 connectivity using the ping and trace-route utility.

If IPv6 connectivity is OK, the issue could be related to the application being used: for example, using a web browser that does not support IPv6 for accessing an IPv6-only website. So in this case, the user can try a different web browser to see if that resolves the issue. In other cases the problem could be related to IPv6 provisioning: for example, the host or GWR not being able to acquire a global unicast IPv6 address to connect to the IPv6 Internet. The host or GWR may be provisioned with a unique-local IPv6 address that is not routable on the IPv6 Internet; therefore, the user would not be able to access a particular IPv6 website or application using this address. Other issues may include problems with DNS server response time, which may cause delay in accessing the website or the DNS server not returning an AAAA record for IPv6, which may cause the connection to time-out if the application does not fall back to using IPv4 in a timely manner.

Other issues could include IPv6 routing problems in the SP network, which could result in the loss of IPv6 connectivity. There could be a firewall or access lists configured on an upstream router that could be blocking IPv6 traffic. In this case the user can try to connect to the Internet using its IPv4 address. If the website or web server being accessed is in dual-stack mode, the user can try connecting over IPv4 to narrow the problem. If the user is trying to access an IPv6-only website or web server, one thing to verify is the client application, to make sure that it is not using IPv4 to make the connection.

The user can also try to capture traffic off the GWR LAN interface by using an application such as Wireshark [21] installed on the host. This can help pinpoint the problem by providing details of what is being transmitted and received on the wire (at the physical layer). Figure 5.35 illustrates an output of packet capture of a failed HTTP connection when a user is trying to access an IPv6-enabled website, but the connection is being reset by the server. Similarly, if the host or GWR is having issues acquiring an IPv6 address, the DHCPv6 or ND messages can be captured using such an application on the host. The packet capture can provide details on where the message exchange is

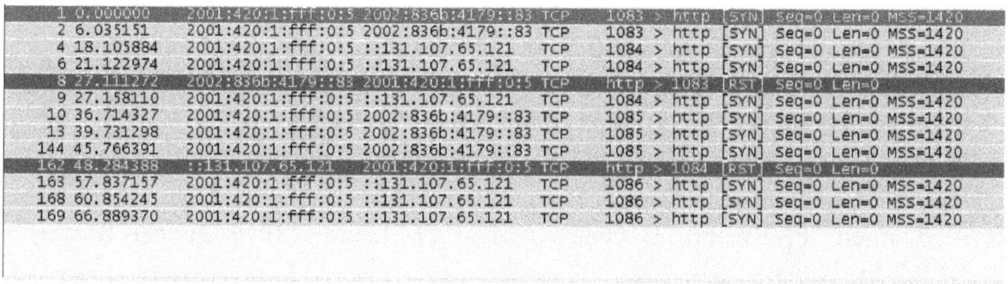

```
   1  0.000000    2001:420:1:fff:0:3  2002:836b:4179::83  TCP    1083 > http  [SYN]  Seq=0 Len=0 MSS=1420
   2  6.035151    2001:420:1:fff:0:5  2002:836b:4179::83  TCP    1083 > http  [SYN]  Seq=0 Len=0 MSS=1420
   4  18.105884   2001:420:1:fff:0:5  ::131.107.65.121    TCP    1084 > http  [SYN]  Seq=0 Len=0 MSS=1420
   6  21.122974   2001:420:1:fff:0:5  ::131.107.65.121    TCP    1084 > http  [SYN]  Seq=0 Len=0 MSS=1420
   8  27.111272   2002:836b:4179::83  2001:420:1:fff:0:5  TCP    http > 1083  [RST]  Seq=0 Len=0
   9  27.158110   2001:420:1:fff:0:5  ::131.107.65.121    TCP    1084 > http  [SYN]  Seq=0 Len=0 MSS=1420
  10  36.714327   2001:420:1:fff:0:5  2002:836b:4179::83  TCP    1085 > http  [SYN]  Seq=0 Len=0 MSS=1420
  13  39.731298   2001:420:1:fff:0:5  2002:836b:4179::83  TCP    1085 > http  [SYN]  Seq=0 Len=0 MSS=1420
 144  45.766391   2001:420:1:fff:0:5  2002:836b:4179::83  TCP    1085 > http  [SYN]  Seq=0 Len=0 MSS=1420
 162  48.284388   ::131.107.65.121    2001:420:1:fff:0:5  TCP    http > 1084  [RST]  Seq=0 Len=0
 163  57.837157   2001:420:1:fff:0:5  ::131.107.65.121    TCP    1086 > http  [SYN]  Seq=0 Len=0 MSS=1420
 168  60.854245   2001:420:1:fff:0:5  ::131.107.65.121    TCP    1086 > http  [SYN]  Seq=0 Len=0 MSS=1420
 169  66.889370   2001:420:1:fff:0:5  ::131.107.65.121    TCP    1086 > http  [SYN]  Seq=0 Len=0 MSS=1420
```

FIGURE 5.35   Packet capture of a failed HTTP connection.

failing and if certain packets are being dropped in the SP network. Refer to Chapter 6 for detailed DHCPv6 debugs on the GWR.

5.8   SUMMARY

Several options are available for deploying IPv6 on hosts and GWR. If the SP provides IPv6 connectivity to the end users and offers IPv6-based services, customers can upgrade their hosts and GWR to dual-stack mode to take advantage of IPv6. If the SP does not offer IPv6 services to end users, customers can tunnel IPv6 traffic over the SP network and connect to the IPv6 Internet. The host and the GWR may need to be upgraded and configured properly to support the tunneling techniques required to enable IPv6 connectivity.

In the end, the decision to enable IPv6 on hosts and GWR depends on customer needs and motivation to connect to the IPv6 Internet. If the SP and IPv6 Internet can offer value-added services to customers (peer-to-peer applications, gaming, high-quality voice and video services), it may strengthen the case for IPv6 deployment in customer premises as well as the SP environment.

REFERENCES

1. T. Narten and R. Draves, "Privacy Extensions for Stateless Address Autoconfiguration in IPv6," RFC3041, January 2001.
2. G. Van de Velde, T. Hain, R. Droms, B. Carpenter, and E. Klein, "Local Network Protection for IPv6," RFC4864, May 2007.
3. S. Thomson, T. Narten, and T. Jinmei, "IPv6 Stateless Address Autoconfiguration", RFC4862, September 2007.
4. G. Malkin, and R. Minnear, "RIPng for IPv6," RFC2080, January 1997.
5. Dibbler, "DHCPv6: Dibbler: A Portable DHCPv6." Available at http://klub.com. pl/dhcpv6/.
6. F. Templin, T. Gleeson, and D. Thaler, "Intra-Site Automatic Tunnel Addressing Protocol," RFC5214, March 2008.

7. C. Huitema, "Teredo: Tunneling IPv6 Over UDP Through Network Address Translations (NATs)," RFC4380, February 2006.

8. Microsoft TechNet, "Teredo Overview." Available at https://www.microsoft.com/technet/network/ipv6/teredo.mspx. Published on January 1, 2003.

9. B. Carpenter, and K. Moore, "Connection of IPv6 Domains via IPv4 Clouds," RFC3056, February 2001.

10. A. Durand, P. Fasano, I. Guardini, and D. Lento, "IPv6 Tunnel Broker," RFC3053, January 2001.

11. S. McFarland, "BRKIPM-2005: Enterprise IPv6 Deployment," Cisco Networkers 2008, January 2008.

12. IPv6 Resource, "Apple MAC OS X IPv6." Available at http://internecine.eu/systems/mac_os_x-ipv6.html. Published on January 2, 2008.

13. Microsoft TechNet, "Netsh Commands for Interface (IPv4 and IPv6)." Available at http://technet2.microsoft.com/windowsserver2008/en/library/29933987-90dc-471c-98aa-04e5fa245bb11033.mspx?mfr = true. Updated on February 8, 2008.

14. Microsoft TechNet, "Routing IPv6 Traffic Over an IPv4 Infrastructure." Available at http://technet2.microsoft.com/WindowsServer/en/library/1512cdf6-fe3b-41de-a5c3-87dbd35d94a41033.mspx?mfr = true. Updated on March 28, 2003.

15. D. Farinacci, T. Li, S. Hanks, D. Meyer, and P. Traina, "Generic Routing Encapsulation (GRE)," RFC2784, March 2000.

16. Sun Microsystems, "IPv6 Administration Guide." Available at http://docsun.cites.uiuc.edu/sun_docs/C/solaris_9/SUNWaadm/IPV6ADMIN/toc.html. Published in April 2003.

17. H. Singh, and W. Beebee, "IPv6 CPE Router Recommendations." Available at http://www.ietf.org/internet-drafts/draft-wbeebee-ipv6-cpe-router-03.txt. Updated on October 30, 2008.

18. The KAME Project, http://www.kame.net/.

19. WIDE v6 Working Group, http://www.v6.wide.ad.jp/.

20. TAHI Project, http://www.tahi.org/.

21. Wireshark, http://www.wireshark.org.

22. DNS Behavior in Vista, Microsoft TechNet, "Domain Name System Client Behavior in Windows Vista," Available at http://technet.microsoft.com/en-us/library/bb727035.aspx. Published on September 18, 2006.

6 Configuring and Troubleshooting IPv6 on Edge Routers

In earlier chapters we discussed several different broadband access technologies deployed in conjunction with IPv6, and covered issues relating to integration as well as solutions and workarounds. This chapter is focused on actual configuration of the SP edge router (ER). As a matter of familiarity and convenience, we base our examples on routers and software from Cisco Systems, Inc. We assume that the Cisco internetwork operating system (Cisco IOS) command line interface (CLI) is familiar to many in our target audience. However, the concepts covered are applicable to other vendor equipment and software that support the IPv6 feature set under consideration. We consider configuration specifics relevant to both upstream and downstream directions of data flow from an SP ER prospective. Upstream means toward the SP core and downstream means toward the SP customer. We would like to reiterate that the focus of this book remains deploying IPv6 in SP broadband access networks. Hence, deployment details, design recommendations, and configuration specifics for SP core networks are beyond the scope of this book and thus are not covered. However, for completeness and to put end-to-end deployment requirements into perspective, we briefly highlight techniques for IPv6 transport over the SP core network. The SP forwards the IPv6 traffic through its Layer3 core in three popular ways: (1) tunneling IPv6 in IPv4, (2) dual-stack networks, and (3) MPLS 6PE and 6VPE.

6.1 IPv6 CONFIGURATION ON THE EDGE ROUTER

In this section we cover steps for configuring the SP ER using Cisco routers and IOS CLI.

6.1.1 Enabling IPv6 on ER

To enable the forwarding of IPv6 traffic on any Cisco router (including a router acting as ER), configure the following command in global configuration mode:

```
Router(config)#ipv6 unicast-routing
```

Deploying IPv6 in Broadband Access Networks, By Adeel Ahmed and Salman Asadullah
Copyright © 2009 John Wiley & Sons, Inc.

Once IPv6 is enabled, the IPv6 stack running on the router assigns a link-local address automatically on any active interface. The link-local address is sufficient for communication between multiple hosts on the same link. A global or unique-local IPv6 address needs to be configured in order to route packets to any other link.

6.1.2   Configuring ER Upstream Interfaces

The ER upstream interfaces could be configured in different ways, depending on how IPv6 packets will be forwarded to the SP core. Subsequently, configuration is covered briefly for three techniques: tunneling IPv6 in IPv4, dual-stack networks, and MPLS 6PE and 6VPE.

6.1.2.1 Tunneling IPv6 in IPv4   As a temporary solution, the SP can use a tunneling mechanism to forward subscriber IPv6 traffic over its core infra-structure. Several tunneling mechanisms were developed specifically to trans-port IPv6 over existing IPv4 infrastructures. Some of these mechanisms have been standardized and their use depends on the existing SP IPv4 network and the IPv6 services and capabilities required. Recommendations regarding optimal approaches for specific scenarios are beyond the scope of this book. Deploying IPv6 using tunneling techniques may require only limited changes to the network, such as software upgrade on tunnel endpoints. This approach has the least impact on the SP network, as only the tunnel endpoints are upgraded to dual-stack routers in order to initiate and terminate the tunnels. However, as the number of tunnel endpoints rises and the amount of IPv6 traffic grows, the tunneling solution will pose scalability and manage-ment challenges. For example, if an SP network has 50 tunnel endpoints, and a nailed point-to-point tunneling mechanism is used, each tunnel endpoint will have to be configured manually for 49 point-to-point tunnels, also known as a full mesh of tunnels. This may be mitigated with an automatic tunneling mechanism such as ISATAP or 6to4. Although automatic tunneling mechanisms require less manual configuration and are more dynamic in operation, generally they do not support IP multicast. If an SP is running into scaling issues with the tunneling approach and cannot provide IPv6 services comparable to IPv4, it may consider scalable solutions such as a dual-stack deployment and 6PE and 6VPE solutions in the case of an MPLS-enabled network.

In this section we cover configuration examples for tunneling mechanisms such as manual tunneling (RFC2893; Figure 6.1) and GRE tunneling. The only difference in the configuration of these two tunnel types is the *tunnel mode command*, which specifies the type of tunnel. The rather simple configuration for enabling tunneling can be illustrated as follows:

- Configure dual-stack tunnel endpoints.
- Configure both IPv4 and IPv6 addresses at each end of the tunnel.

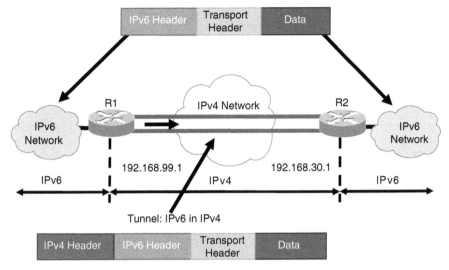

FIGURE 6.1 Manual/GRE tunneling scenarios.

Manual Tunneling (RFC2893) Configuration Example for Upstream Interfaces

```
R1#
interface Tunnel0
 ipv6 enable
 !Enabling IPv6 on Tunnel0 interface.
 ipv6 address 2001:db8:c18:1::3/126
 tunnel source 192.168.99.1
 tunnel destination 192.168.30.1
 tunnel mode ipv6ip
 !Tunneling mode type.
!
R2#
interface Tunnel0
 ipv6 enable
 !Enabling IPv6 on Tunnel0 interface.
 ipv6 address 2001:db8:c18:1::2/126
 tunnel source 192.168.30.1
 tunnel destination 192.168.99.1
 tunnel mode ipv6ip
 !Tunneling mode type.
```

IPv6 over GRE Configuration Example for Upstream Interfaces

```
R1#
interface Tunnel0
```

```
ipv6 enable
!Enabling IPv6 on Tunnel0 interface.
ipv6 address 2001:db8:c18:1::3/126
tunnel source 192.168.99.1
tunnel destination 192.168.30.1
tunnel mode gre ip
!Tunneling mode type.
!
R2#
interface Tunnel0
ipv6 enable
!Enabling IPv6 on Tunnel0 interface.
ipv6 address 2001:db8:c18:1::2/126
tunnel source 192.168.30.1
tunnel destination 192.168.99.1
tunnel mode gre ip
!Tunneling mode type.
```

6.1.2.2 Dual-Stack Networks   Any device with the ability to support IPv6 and IPv4 is referred to as dual-stack. The IPv4-only SP routers can be upgraded to be dual-stack-capable. An IPv6 interior gateway protocol (IGP) such as OSPFv3, ISISv6, or EIGRPv6 may be enabled in a dual-stack network. While upgrading a network to be dual-stack-capable, several network resource issues may emerge which may be worked out easily with proper planning ahead. For example, if the SP is currently running OSPFv2 for IPv4 only and wants to enable OSPFv3 for IPv6, this will essentially place more demand on the routers for memory and processing resources. Also, the introduction of a new protocol will require training for testing and operations staff. It is strongly recommended that the protocol be well tested in a lab environment and in a smaller portion of production network before general deployment throughout the network.

Native IPv4 networks can easily be upgraded to dual-stack mode for supporting IPv6 unicast and multicast traffic, but the added capability may affect IPv4 support operations significantly, due to the additional configuration management responsibilities and possibly the need for installation of new hardware. For dual-stack networks the upstream interfaces will be configured with both IPv4 and IPv6 addresses and the routers enabled with IPv4 and IPv6

FIGURE 6.2   Dual-stack scenario.

routing support (Figure 6.2). In the following example the dual-stack routers are enabled for ISIS routing to support both IPv4 and IPv6.

```
R1#
interface Serial2/0
 ip address 10.1.1.1 255.255.255.0
 ipv6 address 2001:db8:ffff:1::1/64
 ip router isis
 !Enabling ISIS for IPv4.
 ipv6 router isis
 !Enabling ISIS for IPv6.
 isis circuit-type level-2
router isis
 address-family ipv6
 redistribute static
 exit-address family
 net 49.0001.1921.6803.0001.00
!
R2#
interface Serial2/0
 ip address 10.1.1.2 255.255.255.0
 ipv6 address 2001:db8:ffff:1::2/64
  ip router isis
 !Enabling ISIS for IPv4.
  ipv6 router isis
 !Enabling ISIS for IPv6.
  isis circuit-type level-2
router isis
 address-family ipv6
 redistribute static
 exit-address family
 net 49.0002.1921.6803.0002.00
```

6.1.2.3   MPLS 6PE and 6VPE   If the SP is running MPLS in its IPv4 core, it could use 6PE [a term for an IPv6-enabled provider edge (PE) router] to transport IPv6 traffic over an IPv4-only enabled core. As shown in Figure 6.3, 6PE does not support virtual private networks (VPNs) and just provides a mechanism for tunneling IPv6 packets from ingress PE to egress PE routers. MPLS 6PE, also called BGP tunneling, is explained in RFC4798. 6VPE refers to a PE router capable of supporting IPv6 VPNs. As in the case of 6PE, when using 6VPE to support IPv6 transport, the core MPLS network is untouched (no HW/SW upgrade configuration change is needed). Details of the MPLS 6VPE solution are covered in RFC4659. 6VPE solution can be used to provide IPv6-based Layer3 VPN services in a way similar to IPv4-based Layer3 VPN services. Figure 6.4 shows IPv6 VPNs deployed in the same way as IPv4 VPNs.

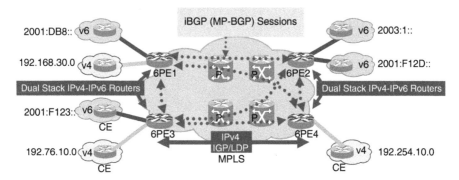

FIGURE 6.3   MPLS 6PE scenario.

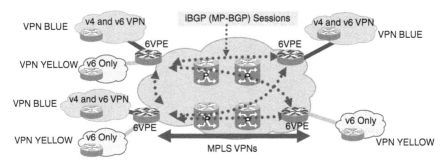

FIGURE 6.4   MPLS 6VPE Scenario.

Some protocol changes made to support IPv6 Layer3 VPN services include the introduction of a new VPNv6 address, MP-BGP VPNv6 address family, VPN IPv6 MP_REACH_NLRI, and encoding of a BGP next hop using an IPv4 mapped IPv6 address. In a nutshell, 6VPE simply adds IPv6 capability to existing IPv4 MPLS VPN services. From the end-user perspective, IPv6-VPN is very similar to the IPv4-VPN service delivery. It is worth noting that 6PE and 6VPE solutions are most suitable for SPs with MPLS networks with MP-BGP and MPLS protocol experience.

The 6PE and 6VPE approaches to providing IPv6 transport are advantageous due to their minimal resource impact on the SP core network. Fewer SP edge devices need to be upgraded and configured to act as dual-stack routers, while the MPLS core continues to be IPv4-only running IPv4 IGP and IPv4 label distribution protocol (LDP). MP-BGP is a common and important component of the 6PE and 6VPE implementation approaches. To reiterate, 6PE and 6VPE functionality is advantageous primarily if MPLS is already deployed in the SP network. IPv6 traffic inherits benefits associated with MPLS, and incremental deployment is also possible. In this scenario, 6PE routers can continue to provide IPv4 and IPv4VPN services while additionally transporting IPv6 packets over the MPLS core. 6VPE routers provide all the functionalities of 6PE routers and IPv6VPN services. IPv6 traffic inherits the MPLS benefits and incremental deployment is possible (i.e., necessitating only upgrading the 6PE routers to 6VPE routers, which have to provide IPv6-VPN services). Each 6VPE router can be connected to sites that are IPv4-only, IPv4VPN-only, IPv4 and IPv6, or IPv4-VPN and IPv6-VPN.

One of the drawbacks of both 6PE and 6VPE solutions is that P routers (since IPv6 is not enabled on them) will not be able to send ICMPv6 messages to support TTL (time to live) expired and trace-route features. This is not a big issue for most networks since these features are currently turned off anyway, to hide the architecture of the network core. However, this limitation raises a valid question regarding IPv6 PMTUD, which relies on ICMPv6 messages. The latter issue can be addressed by MTU tuning at the MPLS PE routers before deploying MPLS 6PE or 6VPE solutions. Cisco also provides a software upgrade for P (provider) routers in MPLS Core to support ICMPv6. At the time of writing this book, a major disadvantage of the 6PE and 6VPE solutions is the inability to natively support IPv6 multicast traffic. This weakness is due to the fact that MPLS does not support multicast natively.

The PE (also known as ER) configuration requirements for integrating 6PE into an MPLS-enabled SP core are discussed below. Refer to Figure 6.3 for the associated MPLS 6PE topology. Even though we refer to 6PE and 6VPE, these could very well be ER routers providing access services to the GWR.

```
6PE1#
ipv6 cef
!
mpls label protocol ldp
!
router bgp 100
 no synchronization
 no bgp default ipv4 unicast
 neighbor 2001:DB8:1::1 remote-as 65014
 neighbor 192.168.99.1 remote-as 100
 neighbor 192.168.99.1 update-source Loopback0
 !
 address-family ipv6
 neighbor 192.168.99.1 activate
 neighbor 192.168.99.1 send-label
 neighbor 2001:DB8:1::1 activate
 redistribute connected
 no synchronization
 exit-address-family
```

In the example above, 2001:DB8:1::1 is the local CE and 192.168.99.1 is the remote 6PE router (6PE2 in Figure 6.3). The **neighbor 192.168.99.1 send-label** command is key to the MPLS 6PE implementation. This command enables

MPLS 6PE functionality by carrying IPv6 prefixes from the ingress, 6PE1, to the egress, 6PE2. A similar configuration will be required at the egress 6PE2 router to perform the same functionality in reverse.

Below are some show commands at the 6PE1 router to verify the 6PE functionality and to demonstrate how routes are being learned. The **show ip route 192.168.99.1** command output shows that 192.168.99.1, which is the IPv4 address of 6PE2, is learned via ISIS (IPv4). This is the address used for iBGP peering with 6PE1.

```
6PE-1#show ip route 192.168.99.1
Routing entry for 192.168.99.1/32
 Known via ''isis'', distance 115, metric 20, type level-2
 * 10.12.0.1, from 192.168.99.1, via FastEthernet1/0
 Route metric is 20, traffic share count is 1
```

The command **show ipv6 route** is used on 6PE1 to see how 2001:F12D::/64 (connected to 6PE2 router) is learned. 2001:F12D::/64 is learned via ::FFFF:192.168.99.1, which is the IPv6 next hop for 6PE1, and it is an IPv4 mapped IPv6 address built from 192.168.99.1.

```
6PE-1#show ipv6 route
B 2001:F12D::/64 [200/0]
 via ::FFFF:192.168.99.1, IPv6-mpls
```

The reader is encouraged to visit www.cisco.com for detailed configuration guide and design papers on MPLS 6PE and 6VPE solutions.

6.1.3   Configuring SP ER Downstream Interfaces

Several features are enabled on the downstream ER interfaces to support IPv6. These functionalities are based on the GWR provisioning requirements, and the ER is expected to provide support based on these requirements. For example, the ER router could behave as a DHCPv6-PD relay router, a DHCPv6-PD server, or a PPP head-end router supporting PPP endpoints. These varying ER functionalities would require some common and some different configuration tasks on the downstream interfaces. In this section we discuss ER configuration requirements to support the following services:

- Configuring IPv6 only or dual-stack ER downstream interfaces
- Configuring SP ER as a DHCPv6-PD relay
- Configuring SP ER as a DHCPv6-PD server
- Configuring SP ER as a DHCPv6-PD server using DUID
- Configuring common PPP-based models

6.1.3.1   IPv6-Only or Dual-Stack SP ER Downstream Interfaces   The downstream SP ER interfaces could be configured for either IPv6 only or dual-stack.

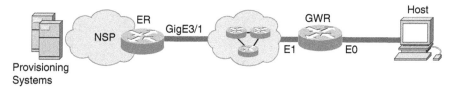

FIGURE 6.5 Access broadband network topology.

If the downstream interface is enabled to support only IPv6, only an IPv6 address is configured on it, and either IPv6 IGP or MP-BGP is used to exchange routing information between the ER and the GWR.

In the following example, interface Gigabit Ethernet 3/1 is enabled for IPv6-only capability, and IPv6 IGP RIPng (RFC2080) is enabled for route exchange between the GWR and the ER. Refer to Figure 6.5.

```
interface GigabitEthernet3/1
ipv6 address 2001:db8:3800:800:0:1:0:1/64
ipv6 enable
ipv6 rip PE_Router enable
!
ipv6 router rip PE_Router
```

The interface-level **ipv6 rip PE_Router enable** command means that the RIPng process is enabled on interface Gigabit Ethernet3/1, by indicating the identifier of RIPng process, PE_Router. For all IPv6 IGPs, the routing process has to be activated at the interface level in addition to the global router-level process. The global command **ipv6 router rip PE_Router** enables the RIPng process on the router.

If the downstream interface is configured for dual-stack capability, it means that both IPv4 and IPv6 addresses are configured on the downstream interface. IPv4 IGP and IPv6 IGP or MP-BGP is used to exchange IPv4 and IPv6 routing information, respectively, between the ER and the GWR. An example of this is shown below, where interface Gigabit Ethernet 3/1 is configured for dual-stack capability. RIP version 2 is configured to exchange IPv4 routes, and RIPng is enabled to exchange IPv6 routes between the ER and the GWR. Note in the configuration below that RIP version 2 is enabled for the entire 10.0.0.0 classful network.

```
interface GigabitEthernet3/1
ipv4 address 10.1.1.1 255.255.255.0
ipv6 address 2001:db8:3800:800:0:1:0:1/64
ipv6 enable
ipv6 rip PE_Router enable
!
ipv6 router rip PE_Router
!
```

```
router rip
version 2
network 10.0.0.0
```

6.1.3.2 *Configuring SP ER as a DHCPv6-PD Relay* In this deployment model, the ER router acts as a DHCPv6-PD relay, forwarding all the messages coming from GWR to the centralized DHCPv6 server (Figure 6.6). The centralized DHCPv6 server needs to be preconfigured in order to decide which IPv6 prefix should be delegated to the GWR and needs to keep track of the list of prefixes delegated. The responses coming from the DHCPv6 server are forwarded to the GWR by the DHCPv6-PD relay router. Below is the configuration needed for an ER acting as a DHCPv6-PD relay, as well as the ND configurations required.

```
hostname ER_Router
!
interface GigabitEthernet3/1
description interface toward the GWR
ipv6 address 2001:420:3800:800:0:1:0:1/64
! Manually configures the address on the interface
GigabitEthernet3/1.
ipv6 nd ra-interval 5
 ! Send an RA message every 5 seconds.
 ipv6 nd prefix default no-advertise
 ! No prefix is advertised.
 ipv6 nd managed-config-flag
 ! M-bit is set.
 ipv6 nd other-config-flag
 ! The O bit is set.
 ! When both M- and O-bits are set in the RA of the ND
 protocol; it means that the ER is telling the GWR to use
 DHCPv6 for address assignment and other configurations as
 well (DNS, etc.) (i.e., stateful DHCPv6).
 ipv6 dhcp relay destination 2001:420:8:1:5::2
 GigabitEthernet0/1
 ! 2001:420:8:1:5::2 is the IPv6 address of the DHCPv6
 server, and the packets coming from GWR are going to be
```

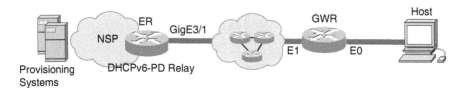

FIGURE 6.6 SP ER acting as a DHCPv6-PD relay.

```
forwarded to this address.
!
interface GigabitEthernet0/1
ip address 10.89.240.235 255.255.255.248
ipv6 address 2001:420:3800:800::12/124
! Manually configures the address on the interface
GigabitEthernet0/1, where the DHCPv6 server is connected.
```

Troubleshooting SP ER Acting as a DHCPv6-PD Relay Once the SP ER is configured for DHCPv6-PD relay functionality; turn on the following debugs on the ER and GWR routers to observe the message exchange. Understanding the messages and negotiations between ER and GWR routers as shown in the debugs is essential for troubleshooting problems.

> debug ipv6 nd
> debug ipv6 dhcp detail
> debug ipv6 dhcp relay

As discussed in earlier chapters, DHCPv6 and DHCPv4 functionalities are similar and are based on a four-way handshake. When the DHCPv6-PD relay is placed in between, it basically forwards messages in both directions (i.e., between the client and the DHCPv6 server). Details of DHCPv6 are discussed in Chapter 2. The following eight messages are seen on the SP ER acting as a DHCPv6-PD relay router and the GWR acting as a client.

1. SOLICIT (CLIENT)
2. RELAY FORWARD W/SOLICIT (RELAY ROUTER)
3. RELAY REPLY W/ADVERTISE (SERVER)
4. ADVERTISE (RELAY ROUTER)
5. REQUEST (CLIENT)
6. RELAY FORWARD W/REQUEST (RELAY ROUTER)
7. RELAY RELPLY W/REPLY (SERVER)
8. REPLY (RELAY ROUTER)

ER Debugs: Initial Address Assignment Request from GWR

```
*Feb 15 21:35:16.946: ICMPv6-ND: Received NS for
FE80::207:EFF:FE03:6E65 on GigE3/1 from ::
! DAD request from the GWR for the link-local address.
*Feb 15 21:35:17.650: ICMPv6-ND: Sending RA to FF02::1 on
GigE3/1
*Feb 15 21:35:17.650: ICMPv6-ND: MTU=1500
```

```
*Feb 15 21:35:17.934: ICMPv6-ND: Received NA for
FE80::207:EFF:FE03:6E65 on GigE3/1 from
FE80::207:EFF:FE03:6E65
```
! GWR assigns the link-local address and sends an NA.

GWR Debugs: Initial ND Negotiations Seen on GWR

```
*Mar 2 02:44:54.349: ICMPv6-ND: Received RA from
FE80::21A:C4FF:FE29:1155 on Ethernet1
*Mar 2 02:44:54.349: ICMPv6-ND: Selected new default
router FE80::21A:C4FF:FE29:1155 on Ethernet1
*Mar 2 02:44:54.353: ICMPv6-ND: checking DHCP
*Mar 2 02:44:54.353: ICMPv6-ND: stateless DHCP
*Mar 2 02:44:54.357: ICMPv6-ND: stateful DHCP
*Mar 2 02:44:54.357: ICMPv6-ND: M bit set; checking prefix
delegation DHCP
*Mar 2 02:44:54.357: ICMPv6-ND: O bit set;
```
*! Since both M- and O-bits are configured on the ER router
interface (Gigabit 3/1 facing the GWR) and this can be
seen in the RA of the ND protocol; it means that the ER is
telling the GWR to use DHCPv6 for address assignment and
other configurations as well (DNS, etc.) (i.e., stateful
DHCPv6).*
```
*Mar 2 02:45:02.709: ICMPv6-ND: Sending NS for
FE80::207:EFF:FE03:6E65 on Ethernet1
```
*! GWR sends a DAD request for a link-local address for its
upstream interface Ethernet1, toward ER.*
```
*Mar 2 02:45:03.709: ICMPv6-ND: DAD:
FE80::207:EFF:FE03:6E65 is unique.
*Mar 2 02:45:03.709: ICMPv6-ND: Sending NA for
FE80::207:EFF:FE03:6E65 on Ethernet1
*Mar 2 02:45:03.709: ICMPv6-ND: Linklocal
FE80::207:EFF:FE03:6E65 on Ethernet1, Up
*Mar 2 02:45:03.717: ICMPv6-ND: Address
FE80::207:EFF:FE03:6E65/10 is up on Ethernet1
*Mar 2 02:45:04.221: ICMPv6-ND: Received RA from
FE80::21A:C4FF:FE29:1155 on Ethernet1
*Mar 2 02:45:04.225: ICMPv6-ND: checking stateless DHCP
*Mar 2 02:45:04.225: ICMPv6-ND: O bit set;
*Mar 2 02:45:06.509: ICMPv6-ND: Prefix Information change
for 2001:420:8:1:7::/80 !
```
! DHCP-PD prefix.
```
*Mar 2 02:45:06.509: ICMPv6-ND: Adding prefix 2001:420:8::/
48 to Ethernet0
*Mar 2 02:45:06.513: ICMPv6-ND: Sending NS for
```

2001:420:8:1:7::1 on Ethernet0
*Mar 2 02:45:06.513: ICMPv6-ND: Prefix Information change
for 2001:420:8:1:6:1:1:EBF1/128
*Mar 2 02:45:06.517: ICMPv6-ND: Adding prefix
2001:420:8:1:6:1:1:EBF1/128 to Ethernet1
*Mar 2 02:45:06.517: ICMPv6-ND: Sending NS for
2001:420:8:1:6:1:1:EBF1 on Ethernet1
*Mar 2 02:45:07.517: ICMPv6-ND: DAD:
2001:420:8:1:6:1:1:EBF1 is unique.
*Mar 2 02:45:07.517: ICMPv6-ND: Sending NA for
2001:420:8:1:6:1:1:EBF1 on Ethernet1
*Mar 2 02:45:07.517: ICMPv6-ND: Address
2001:420:8:1:6:1:1:EBF1/128 is up on Ethernet1
*Mar 2 02:45:07.193: ICMPv6-ND: Request to send RA for
FE80::207:EFF:FE03:6E64
*Mar 2 02:45:07.193: ICMPv6-ND: Sending RA from
FE80::207:EFF:FE03:6E64 to FF02::1 on Ethernet0
*Mar 2 02:45:07.193: ICMPv6-ND: Prefix=2001:420:8:1::/64
onlink autoconfig
*Mar 2 02:45:07.193: ICMPv6-ND: 1209600/604800 (valid/
preferred)
*Mar 2 02:45:07.513: ICMPv6-ND: DAD: 2001:420:8:1:7::1 is
unique.
*Mar 2 02:45:07.513: ICMPv6-ND: Sending NA for
2001:420:8:1:7::1 on Ethernet0
*Mar 2 02:45:07.513: ICMPv6-ND: Address
2001:420:8:1:7::1/80 is up on Ethernet0
*Mar 2 02:45:07.717: ICMPv6-ND: STALE -> DELAY:
FE80::21A:C4FF:FE29:1155
*Mar 2 02:45:10.353: ICMPv6-ND: Received NS for
FE80::207:EFF:FE03:6E65 on Ethernet1 from
FE80::21A:C4FF:FE29:1155
! ND message exchange from the ER to the GWR.
*Mar 2 02:45:10.353: ICMPv6-ND: Sending NA for
FE80::207:EFF:FE03:6E65 on Ether1
! ND message exchange from the GWR to the ER.
*Mar 2 02:45:12.717: ICMPv6-ND: DELAY -> PROBE:
FE80::21A:C4FF:FE29:1155
*Mar 2 02:45:12.717: ICMPv6-ND: Sending NS for
FE80::21A:C4FF:FE29:1155 on Ethernet1
! ND message exchange from the GWR to the ER.
*Mar 2 02:45:12.733: ICMPv6-ND: Received NA for
FE80::21A:C4FF:FE29:1155 on Ethernet1 from
FE80::21A:C4FF:FE29:1155
! ND message exchange from the ER to the GWR.

```
*Mar 2 02:45:12.737: ICMPv6-ND: PROBE -> REACH:
FE80::21A:C4FF:FE29:1155
```

GWR Debugs: ND-SOLICIT

```
*Mar 2 03:39:22.613: IPv6 DHCP: Sending SOLICIT to
FF02::1:2 on Ethernet1
*Mar 2 03:39:22.613: IPv6 DHCP: detailed packet contents
*Mar 2 03:39:22.613: src FE80::207:EFF:FE03:6E65
*Mar 2 03:39:22.613: dst FF02::1:2 (Ethernet1)
```
! GWR sends a SOLICIT message to
All_DHCP_Relay_Agents_and_Servers Address.
```
*Mar 2 03:39:22.613: type SOLICIT(1), xid 16585219
*Mar 2 03:39:22.617: option ELAPSED-TIME(8), len 2
*Mar 2 03:39:22.617: elapsed-time 0
*Mar 2 03:39:22.617: option CLIENTID(1), len 10
*Mar 2 03:39:22.617: 0003000100070E036E65
*Mar 2 03:39:22.617: option IA-NA(3), len 12
*Mar 2 03:39:22.617: IAID 0×00020001, T1 0, T2 0
*Mar 2 03:39:22.617: option IA-PD(25), len 12
*Mar 2 03:39:22.617: IAID 0×00020001, T1 0, T2 0
*Mar 2 03:39:22.621: option ORO(6), len 4
*Mar 2 03:39:22.621: DNS-SERVERS,DOMAIN-LIST
```

ER Debugs: ND-SOLICIT

```
*Feb 15 21:35:19.862: IPv6 DHCP: Received SOLICIT from
FE80::207:EFF:FE03:6E65 on GigE3/1
```
!A SOLICIT message received from the GWR on the interface
GigabitEthernet3/1of the ER.
```
*Feb 15 21:35:19.862: IPv6 DHCP: detailed packet contents
*Feb 15 21:35:19.862: src FE80::207:EFF:FE03:6E65 (GigE3/1)
*Feb 15 21:35:19.862: dst FF02::1:2
*Feb 15 21:35:19.862: type SOLICIT(1), xid 13518535
*Feb 15 21:35:19.862: option ELAPSED-TIME(8), len 2
*Feb 15 21:35:19.862: elapsed-time 0
*Feb 15 21:35:19.862: option CLIENTID(1), len 10
*Feb 15 21:35:19.862: 0003000100070E036E65
*Feb 15 21:35:19.862: option IA-NA(3), len 12
*Feb 15 21:35:19.862: IAID 0×00020001, T1 0, T2 0
*Feb 15 21:35:19.862: option IA-PD(25), len 12
*Feb 15 21:35:19.862: IAID 0×00020001, T1 0, T2 0
*Feb 15 21:35:19.862: option ORO(6), len 4
*Feb 15 21:35:19.862: DNS-SERVERS,DOMAIN-LIST
```

ER Debugs: RELAY-FORWARD with SOLICIT

```
Feb 15 21:35:19.862: IPv6 DHCP_RELAY: Relaying SOLICIT
from FE80::207:EFF:FE03:6E65 on GigE3/1
! ER received a SOLICIT request on the interface
GigabitEthernet3/1from the GWR.
*Feb 15 21:35:19.862: IPv6 DHCP_RELAY: to
2001:420:8:1:5::2 viaGigabitEthernet0/1
*Feb 15 21:35:19.862: IPv6 DHCP: Sending RELAY-FORWARD to
2001:420:8:1:5::2 on GigabitEthernet0/1 next hop
FE80::201:97FF:FE39:2070
! ER forwarding the SOLICIT message to the DHCPv6 server.
*Feb 15 21:35:19.862: IPv6 DHCP: detailed packet contents
*Feb 15 21:35:19.862: src 2001:420:8:1:1::2
*Feb 15 21:35:19.862: dst 2001:420:8:1:5::2
(GigabitEthernet0/1)
*Feb 15 21:35:19.862: type RELAY-FORWARD(12), hop 0
*Feb 15 21:35:19.862: link 2001:420:8:1:6:1:1:1
*Feb 15 21:35:19.862: peer FE80::207:EFF:FE03:6E65
*Feb 15 21:35:19.862: option RELAY-MSG(9), len 64
*Feb 15 21:35:19.862: type SOLICIT(1), xid 13518535
*Feb 15 21:35:19.862: option ELAPSED-TIME(8), len 2
*Feb 15 21:35:19.862: elapsed-time 0
*Feb 15 21:35:19.862: option CLIENTID(1), len 10
*Feb 15 21:35:19.862: 0003000100070E036E65
*Feb 15 21:35:19.862: option IA-NA(3), len 12
*Feb 15 21:35:19.862: IAID 0×00020001, T1 0, T2 0
*Feb 15 21:35:19.862: option IA-PD(25), len 12
*Feb 15 21:35:19.862: IAID 0×00020001, T1 0, T2 0
*Feb 15 21:35:19.862: option ORO(6), len 4
*Feb 15 21:35:19.862: DNS-SERVERS,DOMAIN-LIST
*Feb 15 21:35:19.862: option INTERFACE-ID(18), len 4
*Feb 15 21:35:19.862: 0×00000007
```

ER Debugs: RELAY-REPLY with ADVERTISE

```
*Feb 15 21:35:19.866: IPv6 DHCP: Received RELAY-REPLY from
2001:420:8:1:5::2 on GigE0/1
!The ER received an ADVERTISE from the DHCPv6 server.
*Feb 15 21:35:19.866: IPv6 DHCP: detailed packet contents
*Feb 15 21:35:19.866: src 2001:420:8:1:5::2
(GigabitEthernet0/1)
*Feb 15 21:35:19.866: dst 2001:420:8:1:1::2
*Feb 15 21:35:19.866: type RELAY-REPLY(13), hop 0
```

```
*Feb 15 21:35:19.866: link 2001:420:8:1:6:1:1:1
*Feb 15 21:35:19.866: peer FE80::207:EFF:FE03:6E65
*Feb 15 21:35:19.866: option INTERFACE-ID(18), len 4
*Feb 15 21:35:19.866: 0 × 00000007
*Feb 15 21:35:19.866: option RELAY-MSG(9), len 206
*Feb 15 21:35:19.866: type ADVERTISE(2), xid 13518535
*Feb 15 21:35:19.866: option CLIENTID(1), len 10
*Feb 15 21:35:19.866: 0003000100070E036E65
*Feb 15 21:35:19.866: option SERVERID(2), len 14
*Feb 15 21:35:19.866: 0001000143BF22B6080020E8FAC0
*Feb 15 21:35:19.866: option IA-NA(3), len 40
*Feb 15 21:35:19.866: IAID 0 × 00020001, T1 302400,
T2 483840
*Feb 15 21:35:19.866: option IAADDR(5), len 24
*Feb 15 21:35:19.866: IPv6 address
2001:420:8:1:6:1:1:EBF1
*Feb 15 21:35:19.866: preferred 604800, valid 1209600
*Feb 15 21:35:19.866: option IA-PD(25), len 41
*Feb 15 21:35:19.866: IAID 0 × 00020001, T1 302400,
T2 483840
*Feb 15 21:35:19.866: option IAPREFIX(26), len 25
*Feb 15 21:35:19.866: preferred 604800, valid 1209600,
prefix 2001:420:8::/48
*Feb 15 21:35:19.866: option DNS-SERVERS(23), len 16
*Feb 15 21:35:19.866:
2001:420:3800:801:A00:20FF:FEE5:63E3
*Feb 15 21:35:19.866: option DOMAIN-LIST(24), len 14
*Feb 15 21:35:19.866: v6.cisco.com
```

ER Debugs: ADVERTISE

```
*Feb 15 21:35:19.866: IPv6 DHCP: Sending ADVERTISE to
FE80::207:EFF:FE03:6E65 on GigE3/1
```
! The ER forwards an ADVERTISE message to the GWR.
```
*Feb 15 21:35:19.866: IPv6 DHCP: detailed packet contents
*Feb 15 21:35:19.866: src FE80::21A:C4FF:FE29:1155
*Feb 15 21:35:19.866: dst FE80::207:EFF:FE03:6E65 (GigE3/1)
*Feb 15 21:35:19.866: type ADVERTISE(2), xid 13518535
*Feb 15 21:35:19.866: option CLIENTID(1), len 10
*Feb 15 21:35:19.866: 0003000100070E036E65
*Feb 15 21:35:19.866: option SERVERID(2), len 14
*Feb 15 21:35:19.866: 0001000143BF22B6080020E8FAC0
*Feb 15 21:35:19.866: option IA-NA(3), len 40
*Feb 15 21:35:19.866: IAID 0 × 00020001, T1 302400, T2 483840
```

```
*Feb 15 21:35:19.866: option IAADDR(5), len 24
*Feb 15 21:35:19.866: IPv6 address 2001:420:8:1:6:1:1:EBF1
*Feb 15 21:35:19.866: preferred 604800, valid 1209600
*Feb 15 21:35:19.866: option IA-PD(25), len 41
*Feb 15 21:35:19.866: IAID 0 × 00020001, T1 302400, T2 483840
*Feb 15 21:35:19.866: option IAPREFIX(26), len 25
*Feb 15 21:35:19.866: preferred 604800, valid 1209600, prefix
2001:420:8::/48
*Feb 15 21:35:19.866: option DNS-SERVERS(23), len 16
*Feb 15 21:35:19.866: 2001:420:3800:801:A00:20FF:FEE5:63E3
*Feb 15 21:35:19.866: option DOMAIN-LIST(24), len 14
*Feb 15 21:35:19.866: v6.cisco.com
```

GWR Debugs: ADVERTISE

```
*Mar 2 03:39:22.657: IPv6 DHCP: Received ADVERTISE from
FE80::21A:C4FF:FE29:1155 on Ethernet1
*Mar 2 03:39:22.657: IPv6 DHCP: detailed packet contents
*Mar 2 03:39:22.657: src FE80::21A:C4FF:FE29:1155
(Ethernet1)
! Link-local address of ER.
*Mar 2 03:39:22.657: dst FE80::207:EFF:FE03:6E65
! Link-local address of GWR Ethernet1.
*Mar 2 03:39:22.657: type ADVERTISE(2), xid 16585219
*Mar 2 03:39:22.657: option CLIENTID(1), len 10
*Mar 2 03:39:22.657: 0003000100070E036E65
*Mar 2 03:39:22.661: option SERVERID(2), len 14
*Mar 2 03:39:22.661: 0001000143BF22B6080020E8FAC0
*Mar 2 03:39:22.661: option IA-NA(3), len 40
*Mar 2 03:39:22.661: IAID 0 × 00020001, T1 302400, T2 483840
*Mar 2 03:39:22.661: option IAADDR(5), len 24
*Mar 2 03:39:22.661: IPv6 address 2001:420:8:1:6:1:1:EBF1
*Mar 2 03:39:22.661: preferred 604800, valid 1209600
*Mar 2 03:39:22.665: option IA-PD(25), len 41
*Mar 2 03:39:22.665: IAID 0 × 00020001, T1 302400,
T2 483840
*Mar 2 03:39:22.665: option IAPREFIX(26), len 25
*Mar 2 03:39:22.665: preferred 604800, valid 1209600,
prefix 2001:420:8::/48
*Mar 2 03:39:22.669: option DNS-SERVERS(23), len 16
*Mar 2 03:39:22.669:
2001:420:3800:801:A00:20FF:FEE5:63E3
*Mar 2 03:39:22.669: option DOMAIN-LIST(24), len 14
*Mar 2 03:39:22.669: v6.cisco.com
```

GWR Router Debugs: REQUEST

```
*Mar 2 03:39:23.741: IPv6 DHCP: Sending REQUEST to
FF02::1:2 on Ethernet1
*Mar 2 03:39:23.741: IPv6 DHCP: detailed packet contents
*Mar 2 03:39:23.745: src FE80::207:EFF:FE03:6E65
! Link-local address of GWR Ethernet1.
*Mar 2 03:39:23.745: dst FF02::1:2 (Ethernet1)
! All_DHCP_Relay_Agents_and_Servers Address.
*Mar 2 03:39:23.745: type REQUEST(3), xid 16596644
*Mar 2 03:39:23.745: option ELAPSED-TIME(8), len 2
*Mar 2 03:39:23.745: elapsed-time 0
*Mar 2 03:39:23.745: option CLIENTID(1), len 10
*Mar 2 03:39:23.749: 0003000100070E036E65
*Mar 2 03:39:23.749: option IA-NA(3), len 40
*Mar 2 03:39:23.749: IAID 0 × 00020001, T1 0, T2 0
*Mar 2 03:39:23.749: option IAADDR(5), len 24
*Mar 2 03:39:23.749: IPv6 address 2001:420:8:1:6:1:1:EBF1
*Mar 2 03:39:23.749: preferred 0, valid 0
*Mar 2 03:39:23.749: option IA-PD(25), len 12
*Mar 2 03:39:23.753: IAID 0 × 00020001, T1 0, T2 0
*Mar 2 03:39:23.753: option ORO(6), len 4
*Mar 2 03:39:23.753: DNS-SERVERS,DOMAIN-LIST
*Mar 2 03:39:23.753: option SERVERID(2), len 14
*Mar 2 03:39:23.753: 0001000143BF22B6080020E8FAC0
```

ER Debugs: REQUEST

```
*Feb 15 21:35:20.938: IPv6 DHCP: Received REQUEST from
FE80::207:EFF:FE03:6E65 on GigE3/1
! ER received a REQUEST from the GWR.
*Feb 15 21:35:20.938: IPv6 DHCP: detailed packet contents
*Feb 15 21:35:20.938: src FE80::207:EFF:FE03:6E65 (GigE3/1)
*Feb 15 21:35:20.938: dst FF02::1:2
*Feb 15 21:35:20.938: type REQUEST(3), xid 13530568
*Feb 15 21:35:20.938: option ELAPSED-TIME(8), len 2
*Feb 15 21:35:20.938: elapsed-time 0
*Feb 15 21:35:20.938: option CLIENTID(1), len 10
*Feb 15 21:35:20.938: 0003000100070E036E65
*Feb 15 21:35:20.938: option IA-NA(3), len 40
*Feb 15 21:35:20.938: IAID 0 × 00020001, T1 0, T2 0
*Feb 15 21:35:20.938: option IAADDR(5), len 24
*Feb 15 21:35:20.938: IPv6 address 2001:420:8:1:6:1:1:EBF1
*Feb 15 21:35:20.938: preferred 0, valid 0
*Feb 15 21:35:20.938: option IA-PD(25), len 12
```

```
*Feb 15 21:35:20.938: IAID 0 × 00020001, T1 0, T2 0
*Feb 15 21:35:20.938: option ORO(6), len 4
*Feb 15 21:35:20.938: DNS-SERVERS,DOMAIN-LIST
*Feb 15 21:35:20.938: option SERVERID(2), len 14
*Feb 15 21:35:20.938: 0001000143BF22B6080020E8FAC0
```

ER Debugs: RELAY-FORWARD with REQUEST

```
**Feb 15 21:35:20.938: IPv6 DHCP: Sending RELAY-FORWARD to
2001:420:8:1:5::2 on GigE0/1 next hop
FE80::201:97FF:FE39:2070
```
! ER forwards a REQUEST to the DHCPv6 server again.
```
*Feb 15 21:35:20.938: IPv6 DHCP: detailed packet contents
*Feb 15 21:35:20.938: src 2001:420:8:1:1::2
*Feb 15 21:35:20.938: dst 2001:420:8:1:5::2
(GigabitEthernet0/1)
*Feb 15 21:35:20.938: type RELAY-FORWARD(12), hop 0
*Feb 15 21:35:20.938: link 2001:420:8:1:6:1:1:1
*Feb 15 21:35:20.938: peer FE80::207:EFF:FE03:6E65
*Feb 15 21:35:20.938: option RELAY-MSG(9), len 110
*Feb 15 21:35:20.938: type REQUEST(3), xid 13530568
*Feb 15 21:35:20.938: option ELAPSED-TIME(8), len 2
*Feb 15 21:35:20.938: elapsed-time 0
*Feb 15 21:35:20.938: option CLIENTID(1), len 10
*Feb 15 21:35:20.938: 0003000100070E036E65
*Feb 15 21:35:20.938: option IA-NA(3), len 40
*Feb 15 21:35:20.938: IAID 0 × 00020001, T1 0, T2 0
*Feb 15 21:35:20.938: option IAADDR(5), len 24
*Feb 15 21:35:20.938: IPv6 address
2001:420:8:1:6:1:1:EBF1
*Feb 15 21:35:20.938: preferred 0, valid 0
*Feb 15 21:35:20.938: option IA-PD(25), len 12
*Feb 15 21:35:20.938: IAID 0 × 00020001, T1 0, T2 0
*Feb 15 21:35:20.938: option ORO(6), len 4
*Feb 15 21:35:20.938: DNS-SERVERS,DOMAIN-LIST
*Feb 15 21:35:20.938: option SERVERID(2), len 14
*Feb 15 21:35:20.938: 0001000143BF22B6080020E8FAC0
```

ER Debugs: RELAY-REPLY with REPLY

```
*Feb 15 21:35:20.942: IPv6 DHCP: Received RELAY-REPLY from
2001:420:8:1:5::2 on GigE0/1
```
!The ER received a REPLY from the DHCPv6 server
```
*Feb 15 21:35:20.942: IPv6 DHCP: detailed packet contents
```

```
*Feb 15 21:35:20.942: src 2001:420:8:1:5::2
(GigabitEthernet0/1)
*Feb 15 21:35:20.942: dst 2001:420:8:1:1::2
*Feb 15 21:35:20.942: type RELAY-REPLY(13), hop 0
*Feb 15 21:35:20.942: link 2001:420:8:1:6:1:1:1
*Feb 15 21:35:20.942: peer FE80::207:EFF:FE03:6E65
*Feb 15 21:35:20.942: option INTERFACE-ID(18), len 4
*Feb 15 21:35:20.942: 0 × 00000007
*Feb 15 21:35:20.942: option RELAY-MSG(9), len 206
*Feb 15 21:35:20.942: type REPLY(7), xid 13530568
*Feb 15 21:35:20.942: option CLIENTID(1), len 10
*Feb 15 21:35:20.942: 0003000100070E036E65
*Feb 15 21:35:20.942: option SERVERID(2), len 14
*Feb 15 21:35:20.942: 0001000143BF22B6080020E8FAC0
*Feb 15 21:35:20.942: option IA-NA(3), len 40
*Feb 15 21:35:20.942: IAID 0 × 00020001, T1 302400,
T2 483840
*Feb 15 21:35:20.942: option IAADDR(5), len 24
*Feb 15 21:35:20.942: IPv6 address
2001:420:8:1:6:1:1:EBF1
*Feb 15 21:35:20.942: preferred 604800, valid 1209600
*Feb 15 21:35:20.942: option IA-PD(25), len 41
*Feb 15 21:35:20.942: IAID 0 × 00020001, T1 302400,
T2 483840
*Feb 15 21:35:20.942: option IAPREFIX(26), len 25
*Feb 15 21:35:20.942: preferred 604800, valid 1209600,
prefix 2001:420:8::/48
*Feb 15 21:35:20.942: option DNS-SERVERS(23), len 16
*Feb 15 21:35:20.942:
2001:420:3800:801:A00:20FF:FEE5:63E3
*Feb 15 21:35:20.942: option DOMAIN-LIST(24), len 14
*Feb 15 21:35:20.942: v6.cisco.com
```

ER Debugs: REPLY

```
*Feb 15 21:35:20.942: IPv6 DHCP: Sending REPLY to
FE80::207:EFF:FE03:6E65 on GigE3/1
```
!The ER forwards a REPLY message to the GWR.
```
*Feb 15 21:35:20.942: IPv6 DHCP: detailed packet contents
*Feb 15 21:35:20.942: src FE80::21A:C4FF:FE29:1155
*Feb 15 21:35:20.942: dst FE80::207:EFF:FE03:6E65 (GigE3/1)
*Feb 15 21:35:20.942: type REPLY(7), xid 13530568
*Feb 15 21:35:20.942: option CLIENTID(1), len 10
*Feb 15 21:35:20.942: 0003000100070E036E65
*Feb 15 21:35:20.942: option SERVERID(2), len 14
```

```
*Feb 15 21:35:20.942: 0001000143BF22B6080020E8FAC0
*Feb 15 21:35:20.942: option IA-NA(3), len 40
*Feb 15 21:35:20.942: IAID 0 × 00020001, T1 302400, T2 483840
*Feb 15 21:35:20.942: option IAADDR(5), len 24
*Feb 15 21:35:20.942: IPv6 address 2001:420:8:1:6:1:1:EBF1
*Feb 15 21:35:20.942: preferred 604800, valid 1209600
*Feb 15 21:35:20.942: option IA-PD(25), len 41
*Feb 15 21:35:20.942: IAID 0 × 00020001, T1 302400, T2 483840
*Feb 15 21:35:20.942: option IAPREFIX(26), len 25
*Feb 15 21:35:20.942: preferred 604800, valid 1209600, prefix
2001:420:8::/48
! DHCPv6-PD prefix.
*Feb 15 21:35:20.946: option DNS-SERVERS(23), len 16
*Feb 15 21:35:20.946: 2001:420:3800:801:A00:20FF:FEE5:63E3
*Feb 15 21:35:20.946: option DOMAIN-LIST(24), len 14
*Feb 15 21:35:20.946: v6.cisco.com
```

GWR Debugs: REPLY

```
*Mar 2 03:39:23.797: IPv6 DHCP: Received REPLY from
FE80::21A:C4FF:FE29:1155 on Ethernet1
*Mar 2 03:39:23.797: IPv6 DHCP: detailed packet contents
*Mar 2 03:39:23.797: src FE80::21A:C4FF:FE29:1155
(Ethernet1)
! Link-local address of ER.
*Mar 2 03:39:23.797: dst FE80::207:EFF:FE03:6E65
! Link-local address of GWR Ethernet1.
*Mar 2 03:39:23.801: type REPLY(7), xid 16596644
*Mar 2 03:39:23.801: option CLIENTID(1), len 10
*Mar 2 03:39:23.801: 0003000100070E036E65
*Mar 2 03:39:23.801: option SERVERID(2), len 14
*Mar 2 03:39:23.801: 0001000143BF22B6080020E8FAC0
*Mar 2 03:39:23.801: option IA-NA(3), len 40
*Mar 2 03:39:23.801: IAID 0 × 00020001, T1 302400, T2 483840
*Mar 2 03:39:23.801: option IAADDR(5), len 24
*Mar 2 03:39:23.805: IPv6 address 2001:420:8:1:6:1:1:EBF1
*Mar 2 03:39:23.805: preferred 604800, valid 1209600
*Mar 2 03:39:23.805: option IA-PD(25), len 41
*Mar 2 03:39:23.805: IAID 0 × 00020001, T1 302400, T2 483840
*Mar 2 03:39:23.805: option IAPREFIX(26), len 25
*Mar 2 03:39:23.805: preferred 604800, valid 1209600,
prefix 2001:420:8::/48
*Mar 2 03:39:23.809: option DNS-SERVERS(23), len 16
*Mar 2 03:39:23.809:
2001:420:3800:801:A00:20FF:FEE5:63E3
```

```
*Mar 2 03:39:23.809: option DOMAIN-LIST(24), len 14
*Mar 2 03:39:23.809: v6.cisco.com
```

ER Debugs: ND

```
*Feb 15 21:35:20.970: ICMPv6-ND: Received NS for
2001:420:8:1:6:1:1:EBF1 on GigE3/1 from ::
```
!A DAD Request from the GWR.
```
*Feb 15 21:35:21.490: ICMPv6-ND: Sending RA to FF02::1 on
GigE3/1
*Feb 15 21:35:21.490: ICMPv6-ND: MTU=1500
*Feb 15 21:35:21.974: ICMPv6-ND: Received NA for
2001:420:8:1:6:1:1:EBF1 on GigE3/1 from
2001:420:8:1:6:1:1:EBF1
```
!The GWR assigns an address and the ER receives an NA from
the GWR.
```
*Feb 15 21:35:24.866: ICMPv6-ND: DELAY -> PROBE:
FE80::207:EFF:FE03:6E65
*Feb 15 21:35:24.866: ICMPv6-ND: Sending NS for
FE80::207:EFF:FE03:6E65 on GigE3/1
```
!The ER sends an NS to the GWR.
```
*Feb 15 21:35:24.878: ICMPv6-ND: Received NA for
FE80::207:EFF:FE03:6E65 on GigE3/1 from
FE80::207:EFF:FE03:6E65
```
!The GWR responds with an NA to the ER.
```
*Feb 15 21:35:24.878: ICMPv6-ND: PROBE -> REACH:
FE80::207:EFF:FE03:6E65
*Feb 15 21:35:26.102: ICMPv6-ND: Sending RA to FF02::1 on
GigE3/1
*Feb 15 21:35:26.102: ICMPv6-ND: MTU=1500
*Feb 15 21:35:28.942: ICMPv6-ND: Received NS for
FE80::21A:C4FF:FE29:1155 on GigE3/1 from
FE80::207:EFF:FE03:6E65
```
!The ER receives an NS from the GWR for its link-local.
```
*Feb 15 21:35:28.942: ICMPv6-ND: Sending NA for
FE80::21A:C4FF:FE29:1155 on GigE3/1
```
!The ER sends an NA to the GWR.
```
*Feb 15 21:35:30.302: ICMPv6-ND: Sending RA to FF02::1 on
GigE3/1
*Feb 15 21:35:30.302: ICMPv6-ND: MTU=1500
```

Verifying IPv6 Addresses Assigned to GWR   Note the IPv6 address assigned via DHCPv6 on the interface Ethernet1 and Ethernet0 on the GWR router. Ethernet1 of the GWR router is the upstream interface facing toward the ER router. Ethernet0 is the downstream interface; facing toward the host

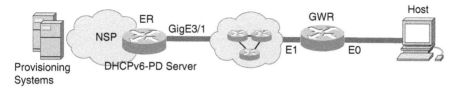

FIGURE 6.7   SP ER acting as a DHCPv6-PD server.

connected to the GWR router. In this scenario, the GWR router is acting as a DHCPv6 client to the ER router and acting as a DHCPv6 server for the host connected to it. Refer to Figure 6.7.

```
GWR Router#show ipv6 interface ethernet1

Ethernet1 is up, line protocol is up
 IPv6 is enabled, link-local address is
FE80::207:EFF:FE03:6E65
 No Virtual link-local address(es):
 Global unicast address(es):
 2001:420:8:1:6:1:1:EBF1, subnet is
2001:420:8:1:6:1:1:EBF1/128 [CAL/PRE]
 ! Address assigned by DHCPv6.
 valid lifetime 1121384 preferred lifetime 516584

GWR Router#show ipv6 interface ethernet0
Ethernet0 is up, line protocol is up
 IPv6 is enabled, link-local address is
FE80::207:EFF:FE03:6E64
 No Virtual link-local address(es):
 Global unicast address(es):
 2001:420:8:1:7::1, subnet is 2001:420:8:1::/64 [CAL/PRE]
 ! Address assigned using DHCP-PD.
 valid lifetime 1121385 preferred lifetime 516585
```

6.1.3.3   Configuring SP ER as a DHCPv6-PD Server   In this deployment model, the SP ER also acts as a DHCPv6-PD server in the absence of a central DHCPv6 server. In this case, all the messages coming from the GWR are processed at the ER router, and responses are sent to the GWR. The ER router needs to be preconfigured in order to decide which IPv6 prefix should be delegated to the GWR, and needs to keep track of the list of prefixes delegated. The configuration needed on the SP ER acting as a DHCPv6-PD server is

```
hostname ER_Router
!
interface GigabitEthernet3/1
```

```
description interface toward the GWR
ipv6 dhcp server RCDN
!
ipv6 dhcp pool RCDN
prefix-delegation pool FOO
!
ipv6 local pool FOO 2001:420:8::/48 48
!
```

Troubleshooting SP ER Acting as a DHCPv6-PD Server   You can enable debugs to observe message exchanges between the ER and the GWR after configuring the SP ER router as a DHCPv6-PD server. The following debug outputs will be useful in troubleshooting and isolating problems.

  debug ipv6 nd
  debug ipv6 dhcp detail
  debug ipv6 dhcp relay

As discussed in earlier chapters, DHCPv6 and DHCPv4 functionalities are similar and based on a four-way hand shake. The following four messages are seen on the ER performing the DHCPv6-PD server functionalities and the GWR acting as a DHCPv6 client:

  1. SOLICIT
  2. ADVERTISE
  3. REQUEST
  4. REPLY

ER Debugs: Initial Address Assignment Request from GWR

```
*Feb 15 21:35:16.946: ICMPv6-ND: Received NS for
FE80::207:EFF:FE03:6E65 on GigE3/1 from ::
! A DAD request from the GWR for the Link-local Address.
*Feb 15 21:35:17.650: ICMPv6-ND: Sending RA to FF02::1 on
GigE3/1
*Feb 15 21:35:17.650: ICMPv6-ND: MTU=1500
*Feb 15 21:35:17.934: ICMPv6-ND: Received NA for
FE80::207:EFF:FE03:6E65 on GigE3/1 from
FE80::207:EFF:FE03:6E65
!The GWR assigns a link-local address and sends an NA.
```

GWR Debugs: Initial ND Negotiations Seen on GWR

```
*Mar 2 02:44:54.349: ICMPv6-ND: Received RA from
FE80::21A:C4FF:FE29:1155 on Ethernet1
```

```
*Mar 2 02:44:54.349: ICMPv6-ND: Selected new default
router FE80::21A:C4FF:FE29:1155 on Ethernet1
*Mar 2 02:44:54.353: ICMPv6-ND: checking DHCP
*Mar 2 02:44:54.353: ICMPv6-ND: stateless DHCP
*Mar 2 02:44:54.357: ICMPv6-ND: stateful DHCP
*Mar 2 02:44:54.357: ICMPv6-ND: M bit set; checking prefix
delegation DHCP
*Mar 2 02:44:54.357: ICMPv6-ND: O bit set;
```
*! Since both M- and O-bits are configured on the ER interface
(Gigabit 3/1 facing the GWR) and this can be seen in the RA
of the ND protocol, it means that the ER is telling the GWR
to use DHCPv6 for address assignment and other
configuration's as well (DNS, etc.) (i.e., Stateful
DHCPv6).*
```
*Mar 2 02:45:02.709: ICMPv6-ND: Sending NS for
FE80::207:EFF:FE03:6E65 on Ethernet1
```
*!The GWR sends a DAD request for the link-local address for
its upstream interface Ethernet1, toward the ER.*
```
*Mar 2 02:45:03.709: ICMPv6-ND: DAD:
FE80::207:EFF:FE03:6E65 is unique.
*Mar 2 02:45:03.709: ICMPv6-ND: Sending NA for
FE80::207:EFF:FE03:6E65 on Ethernet1
*Mar 2 02:45:03.709: ICMPv6-ND: Linklocal
FE80::207:EFF:FE03:6E65 on Ethernet1, Up
*Mar 2 02:45:03.717: ICMPv6-ND: Address
FE80::207:EFF:FE03:6E65/10 is up on Ethernet1
*Mar 2 02:45:04.221: ICMPv6-ND: Received RA from
FE80::21A:C4FF:FE29:1155 on Ethernet1
*Mar 2 02:45:04.225: ICMPv6-ND: checking stateless DHCP
*Mar 2 02:45:04.225: ICMPv6-ND: O bit set;
*Mar 2 02:45:06.509: ICMPv6-ND: Prefix Information change
for 2001:420:8:1:7::/80
```
! DHCP-PD prefix.
```
*Mar 2 02:45:06.509: ICMPv6-ND: Adding prefix 2001:420:8::
/48 to Ethernet0
*Mar 2 02:45:06.513: ICMPv6-ND: Sending NS for
2001:420:8:1:7::1 on Ethernet0
*Mar 2 02:45:06.513: ICMPv6-ND: Prefix Information change
for 2001:420:8:1:6:1:1:EBF1/128
*Mar 2 02:45:06.517: ICMPv6-ND: Adding prefix
2001:420:8:1:6:1:1:EBF1/128 to Ethernet1
*Mar 2 02:45:06.517: ICMPv6-ND: Sending NS for
2001:420:8:1:6:1:1:EBF1 on Ethernet1
*Mar 2 02:45:07.517: ICMPv6-ND: DAD:
2001:420:8:1:6:1:1:EBF1 is unique.
```

```
*Mar 2 02:45:07.517: ICMPv6-ND: Sending NA for
2001:420:8:1:6:1:1:EBF1 on Ethernet1
*Mar 2 02:45:07.517: ICMPv6-ND: Address
2001:420:8:1:6:1:1:EBF1/128 is up on Ethernet1
*Mar 2 02:45:07.193: ICMPv6-ND: Request to send RA for
FE80::207:EFF:FE03:6E64
*Mar 2 02:45:07.193: ICMPv6-ND: Sending RA from
FE80::207:EFF:FE03:6E64 to FF02::1 on Ethernet0
*Mar 2 02:45:07.193: ICMPv6-ND: Prefix=2001:420:8:1::/64
onlink autoconfig
*Mar 2 02:45:07.193: ICMPv6-ND: 1209600/604800 (valid/
preferred)
*Mar 2 02:45:07.513: ICMPv6-ND: DAD: 2001:420:8:1:7::1
is unique.
*Mar 2 02:45:07.513: ICMPv6-ND: Sending NA for
2001:420:8:1:7::1 on Ethernet0
*Mar 2 02:45:07.513: ICMPv6-ND: Address
2001:420:8:1:7::1/80 is up on Ethernet0
*Mar 2 02:45:07.717: ICMPv6-ND: STALE -> DELAY:
FE80::21A:C4FF:FE29:1155
*Mar 2 02:45:10.353: ICMPv6-ND: Received NS for
FE80::207:EFF:FE03:6E65 on Ethernet1 from
FE80::21A:C4FF:FE29:1155
```
! ND message exchange from the ER to the GWR.
```
*Mar 2 02:45:10.353: ICMPv6-ND: Sending NA for
FE80::207:EFF:FE03:6E65 on Ethernet1
```
! ND message exchange from the GWR to the ER.
```
*Mar 2 02:45:12.717: ICMPv6-ND: DELAY -> PROBE:
FE80::21A:C4FF:FE29:1155
*Mar 2 02:45:12.717: ICMPv6-ND: Sending NS for
FE80::21A:C4FF:FE29:1155 on Ethernet1
```
! ND message exchange from the GWR the to ER.
```
*Mar 2 02:45:12.733: ICMPv6-ND: Received NA for
FE80::21A:C4FF:FE29:1155 on Ethernet1 from
FE80::21A:C4FF:FE29:1155
```
! ND message exchange from the ER to the GWR.
```
*Mar 2 02:45:12.737: ICMPv6-ND: PROBE -> REACH:
FE80::21A:C4FF:FE29:1155
```

GWR Debugs: SOLICIT

```
*Mar 2 03:39:22.613: IPv6 DHCP: Sending SOLICIT to
FF02::1:2 on Ethernet1
*Mar 2 03:39:22.613: IPv6 DHCP: detailed packet contents
*Mar 2 03:39:22.613: src FE80::207:EFF:FE03:6E65
```

```
*Mar 2 03:39:22.613: dst FF02::1:2 (Ethernet1)
```
!*The GWR sends a SOLICIT message to*
All_DHCP_Relay_Agents_and_Servers Address.
```
*Mar 2 03:39:22.613: type SOLICIT(1), xid 16585219
*Mar 2 03:39:22.617: option ELAPSED-TIME(8), len 2
*Mar 2 03:39:22.617: elapsed-time 0
*Mar 2 03:39:22.617: option CLIENTID(1), len 10
*Mar 2 03:39:22.617: 0003000100070E036E65
*Mar 2 03:39:22.617: option IA-NA(3), len 12
*Mar 2 03:39:22.617: IAID 0 × 00020001, T1 0, T2 0
*Mar 2 03:39:22.617: option IA-PD(25), len 12
*Mar 2 03:39:22.617: IAID 0 × 00020001, T1 0, T2 0
*Mar 2 03:39:22.621: option ORO(6), len 4
*Mar 2 03:39:22.621: DNS-SERVERS,DOMAIN-LIST
```

ER Debugs: SOLICIT

```
*Feb 15 21:35:19.862: IPv6 DHCP: Received SOLICIT from
FE80::207:EFF:FE03:6E65 on GigE3/1
```
!*A SOLICIT message received from the GWR on the interface*
GigabitEthernet3/1.
```
*Feb 15 21:35:19.862: IPv6 DHCP: detailed packet contents
*Feb 15 21:35:19.862: src FE80::207:EFF:FE03:6E65 (GigE3/1)
*Feb 15 21:35:19.862: dst FF02::1:2
*Feb 15 21:35:19.862: type SOLICIT(1), xid 13518535
*Feb 15 21:35:19.862: option ELAPSED-TIME(8), len 2
*Feb 15 21:35:19.862: elapsed-time 0
*Feb 15 21:35:19.862: option CLIENTID(1), len 10
*Feb 15 21:35:19.862: 0003000100070E036E65
*Feb 15 21:35:19.862: option IA-NA(3), len 12
*Feb 15 21:35:19.862: IAID 0 × 00020001, T1 0, T2 0
*Feb 15 21:35:19.862: option IA-PD(25), len 12
*Feb 15 21:35:19.862: IAID 0 × 00020001, T1 0, T2 0
*Feb 15 21:35:19.862: option ORO(6), len 4
*Feb 15 21:35:19.862: DNS-SERVERS,DOMAIN-LIST
```

ER Debugs: ADVERTISE

```
*Feb 15 21:35:19.866: IPv6 DHCP: Sending ADVERTISE to
FE80::207:EFF:FE03:6E65 on GigE3/1
```
!*The ER sends an ADVERTISE message to the GWR.*
```
*Feb 15 21:35:19.866: IPv6 DHCP: detailed packet contents
*Feb 15 21:35:19.866: src FE80::21A:C4FF:FE29:1155
*Feb 15 21:35:19.866: dst FE80::207:EFF:FE03:6E65 (GigE3/1)
*Feb 15 21:35:19.866: type ADVERTISE(2), xid 13518535
```

```
*Feb 15 21:35:19.866: option CLIENTID(1), len 10
*Feb 15 21:35:19.866: 0003000100070E036E65
*Feb 15 21:35:19.866: option SERVERID(2), len 14
*Feb 15 21:35:19.866: 0001000143BF22B6080020E8FAC0
*Feb 15 21:35:19.866: option IA-NA(3), len 40
*Feb 15 21:35:19.866: IAID 0×00020001, T1 302400, T2 483840
*Feb 15 21:35:19.866: option IAADDR(5), len 24
*Feb 15 21:35:19.866: IPv6 address 2001:420:8:1:6:1:1:EBF1
*Feb 15 21:35:19.866: preferred 604800, valid 1209600
*Feb 15 21:35:19.866: option IA-PD(25), len 41
*Feb 15 21:35:19.866: IAID 0×00020001, T1 302400, T2 483840
*Feb 15 21:35:19.866: option IAPREFIX(26), len 25
*Feb 15 21:35:19.866: preferred 604800, valid 1209600, prefix
2001:420:8::/48
*Feb 15 21:35:19.866: option DNS-SERVERS(23), len 16
*Feb 15 21:35:19.866: 2001:420:3800:801:A00:20FF:FEE5:63E3
*Feb 15 21:35:19.866: option DOMAIN-LIST(24), len 14
*Feb 15 21:35:19.866: v6.cisco.com
```

GWR Debugs: ADVERTISE

```
*Mar 2 03:39:22.657: IPv6 DHCP: Received ADVERTISE from
FE80::21A:C4FF:FE29:1155 on Ethernet1
*Mar 2 03:39:22.657: IPv6 DHCP: detailed packet contents
*Mar 2 03:39:22.657: src FE80::21A:C4FF:FE29:1155
(Ethernet1)
! Link-local Address of ER.
*Mar 2 03:39:22.657: dst FE80::207:EFF:FE03:6E65
! Link-local Address of GWR Ethernet1.
*Mar 2 03:39:22.657: type ADVERTISE(2), xid 16585219
*Mar 2 03:39:22.657: option CLIENTID(1), len 10
*Mar 2 03:39:22.657: 0003000100070E036E65
*Mar 2 03:39:22.661: option SERVERID(2), len 14
*Mar 2 03:39:22.661: 0001000143BF22B6080020E8FAC0
*Mar 2 03:39:22.661: option IA-NA(3), len 40
*Mar 2 03:39:22.661: IAID 0×00020001, T1 302400, T2 483840
*Mar 2 03:39:22.661: option IAADDR(5), len 24
*Mar 2 03:39:22.661: IPv6 address 2001:420:8:1:6:1:1:EBF1
*Mar 2 03:39:22.661: preferred 604800, valid 1209600
*Mar 2 03:39:22.665: option IA-PD(25), len 41
*Mar 2 03:39:22.665: IAID 0×00020001, T1 302400, T2 483840
*Mar 2 03:39:22.665: option IAPREFIX(26), len 25
*Mar 2 03:39:22.665: preferred 604800, valid 1209600,
prefix 2001:420:8::/48
*Mar 2 03:39:22.669: option DNS-SERVERS(23), len 16
```

```
*Mar 2 03:39:22.669:
2001:420:3800:801:A00:20FF:FEE5:63E3
*Mar 2 03:39:22.669: option DOMAIN-LIST(24), len 14
*Mar 2 03:39:22.669: v6.cisco.com
```

GWR Debugs: REQUEST

```
*Mar 2 03:39:23.741: IPv6 DHCP: Sending REQUEST to
FF02::1:2 on Ethernet1
*Mar 2 03:39:23.741: IPv6 DHCP: detailed packet contents
*Mar 2 03:39:23.745: src FE80::207:EFF:FE03:6E65
! Link-local Address of GWR Ethernet1.
*Mar 2 03:39:23.745: dst FF02::1:2 (Ethernet1)
!The GWR router sends a REQUEST to the ER on
All_DHCP_Relay_Agents_and_Servers Address.
*Mar 2 03:39:23.745: type REQUEST(3), xid 16596644
*Mar 2 03:39:23.745: option ELAPSED-TIME(8), len 2
*Mar 2 03:39:23.745: elapsed-time 0
*Mar 2 03:39:23.745: option CLIENTID(1), len 10
*Mar 2 03:39:23.749: 0003000100070E036E65
*Mar 2 03:39:23.749: option IA-NA(3), len 40
*Mar 2 03:39:23.749: IAID 0×00020001, T1 0, T2 0
*Mar 2 03:39:23.749: option IAADDR(5), len 24
*Mar 2 03:39:23.749: IPv6 address 2001:420:8:1:6:1:1:EBF1
*Mar 2 03:39:23.749: preferred 0, valid 0
*Mar 2 03:39:23.749: option IA-PD(25), len 12
*Mar 2 03:39:23.753: IAID 0×00020001, T1 0, T2 0
*Mar 2 03:39:23.753: option ORO(6), len 4
*Mar 2 03:39:23.753: DNS-SERVERS,DOMAIN-LIST
*Mar 2 03:39:23.753: option SERVERID(2), len 14
*Mar 2 03:39:23.753: 0001000143BF22B6080020E8FAC0
```

ER Debugs: REQUEST

```
*Feb 15 21:35:20.938: IPv6 DHCP: Received REQUEST from
FE80::207:EFF:FE03:6E65 on GigE3/1
!The ER received a REQUEST from the GWR.
*Feb 15 21:35:20.938: IPv6 DHCP: detailed packet contents
*Feb 15 21:35:20.938: src FE80::207:EFF:FE03:6E65 (GigE3/1)
*Feb 15 21:35:20.938: dst FF02::1:2
*Feb 15 21:35:20.938: type REQUEST(3), xid 13530568
*Feb 15 21:35:20.938: option ELAPSED-TIME(8), len 2
*Feb 15 21:35:20.938: elapsed-time 0
*Feb 15 21:35:20.938: option CLIENTID(1), len 10
*Feb 15 21:35:20.938: 0003000100070E036E65
```

```
*Feb 15 21:35:20.938: option IA-NA(3), len 40
*Feb 15 21:35:20.938: IAID 0 × 00020001, T1 0, T2 0
*Feb 15 21:35:20.938: option IAADDR(5), len 24
*Feb 15 21:35:20.938: IPv6 address 2001:420:8:1:6:1:1:EBF1
*Feb 15 21:35:20.938: preferred 0, valid 0
*Feb 15 21:35:20.938: option IA-PD(25), len 12
*Feb 15 21:35:20.938: IAID 0 × 00020001, T1 0, T2 0
*Feb 15 21:35:20.938: option ORO(6), len 4
*Feb 15 21:35:20.938: DNS-SERVERS,DOMAIN-LIST
*Feb 15 21:35:20.938: option SERVERID(2), len 14
*Feb 15 21:35:20.938: 0001000143BF22B6080020E8FAC0
```

ER Debugs: REPLY

```
*Feb 15 21:35:20.942: IPv6 DHCP: Sending REPLY to
FE80::207:EFF:FE03:6E65 on GigE3/1
```
!The ER sends a REPLY message to the GWR.
```
*Feb 15 21:35:20.942: IPv6 DHCP: detailed packet contents
*Feb 15 21:35:20.942: src FE80::21A:C4FF:FE29:1155
*Feb 15 21:35:20.942: dst FE80::207:EFF:FE03:6E65 (GigE3/1)
*Feb 15 21:35:20.942: type REPLY(7), xid 13530568
*Feb 15 21:35:20.942: option CLIENTID(1), len 10
*Feb 15 21:35:20.942: 0003000100070E036E65
*Feb 15 21:35:20.942: option SERVERID(2), len 14
*Feb 15 21:35:20.942: 0001000143BF22B6080020E8FAC0
*Feb 15 21:35:20.942: option IA-NA(3), len 40
*Feb 15 21:35:20.942: IAID 0 × 00020001, T1 302400, T2 483840
*Feb 15 21:35:20.942: option IAADDR(5), len 24
*Feb 15 21:35:20.942: IPv6 address 2001:420:8:1:6:1:1:EBF1
*Feb 15 21:35:20.942: preferred 604800, valid 1209600
*Feb 15 21:35:20.942: option IA-PD(25), len 41
*Feb 15 21:35:20.942: IAID 0 × 00020001, T1 302400, T2 483840
*Feb 15 21:35:20.942: option IAPREFIX(26), len 25
*Feb 15 21:35:20.942: preferred 604800, valid 1209600, prefix
2001:420:8::/48
```
! DHCPv6-PD prefix.
```
*Feb 15 21:35:20.946: option DNS-SERVERS(23), len 16
*Feb 15 21:35:20.946: 2001:420:3800:801:A00:20FF:FEE5:63E3
*Feb 15 21:35:20.946: option DOMAIN-LIST(24), len 14
*Feb 15 21:35:20.946: v6.cisco.com
```

GWR Router Debugs: REPLY

```
*Mar 2 03:39:23.797: IPv6 DHCP: Received REPLY from
FE80::21A:C4FF:FE29:1155 on Ethernet1
```

```
*Mar 2 03:39:23.797: IPv6 DHCP: detailed packet contents
*Mar 2 03:39:23.797: src FE80::21A:C4FF:FE29:1155
(Ethernet1)
```
! Link-local address of ER.
```
*Mar 2 03:39:23.797: dst FE80::207:EFF:FE03:6E65
```
! Link-local address of GWR Ethernet1.
```
*Mar 2 03:39:23.801: type REPLY(7), xid 16596644
*Mar 2 03:39:23.801: option CLIENTID(1), len 10
*Mar 2 03:39:23.801: 0003000100070E036E65
*Mar 2 03:39:23.801: option SERVERID(2), len 14
*Mar 2 03:39:23.801: 0001000143BF22B6080020E8FAC0
*Mar 2 03:39:23.801: option IA-NA(3), len 40
*Mar 2 03:39:23.801: IAID 0 × 00020001, T1 302400, T2 483840
*Mar 2 03:39:23.801: option IAADDR(5), len 24
*Mar 2 03:39:23.805: IPv6 address 2001:420:8:1:6:1:1:EBF1
*Mar 2 03:39:23.805: preferred 604800, valid 1209600
*Mar 2 03:39:23.805: option IA-PD(25), len 41
*Mar 2 03:39:23.805: IAID 0 × 00020001, T1 302400, T2 483840
*Mar 2 03:39:23.805: option IAPREFIX(26), len 25
*Mar 2 03:39:23.805: preferred 604800, valid 1209600,
prefix 2001:420:8::/48
*Mar 2 03:39:23.809: option DNS-SERVERS(23), len 16
*Mar 2 03:39:23.809:
2001:420:3800:801:A00:20FF:FEE5:63E3
*Mar 2 03:39:23.809: option DOMAIN-LIST(24), len 14
*Mar 2 03:39:23.809: v6.cisco.com
```

ER Debugs: ND

```
*Feb 15 21:35:20.970: ICMPv6-ND: Received NS for
2001:420:8:1:6:1:1:EBF1 on GigE3/1 from ::
```
!A DAD request from the GWR.
```
*Feb 15 21:35:21.490: ICMPv6-ND: Sending RA to FF02::1
on GigE3/1
*Feb 15 21:35:21.490: ICMPv6-ND: MTU=1500
*Feb 15 21:35:21.974: ICMPv6-ND: Received NA for
2001:420:8:1:6:1:1:EBF1 on GigE3/1 from
2001:420:8:1:6:1:1:EBF1
```
!The GWR assigns an address and ER receives an NA from the GWR.
```
*Feb 15 21:35:24.866: ICMPv6-ND: DELAY -> PROBE:
FE80::207:EFF:FE03:6E65
*Feb 15 21:35:24.866: ICMPv6-ND: Sending NS for
FE80::207:EFF:FE03:6E65 on GigE3/1
```
!The ER sends an NS to the GWR.

```
*Feb 15 21:35:24.878: ICMPv6-ND: Received NA for
FE80::207:EFF:FE03:6E65 on GigE3/1 from
FE80::207:EFF:FE03:6E65
!The GWR responds with an NA to the ER.
*Feb 15 21:35:24.878: ICMPv6-ND: PROBE -> REACH:
FE80::207:EFF:FE03:6E65
*Feb 15 21:35:26.102: ICMPv6-ND: Sending RA to FF02::1
on GigE3/1
*Feb 15 21:35:26.102: ICMPv6-ND: MTU=1500
*Feb 15 21:35:28.942: ICMPv6-ND: Received NS for
FE80::21A:C4FF:FE29:1155 on GigE3/1 from
FE80::207:EFF:FE03:6E65
!The ER receives an NS from the GWR for its link-local.
*Feb 15 21:35:28.942: ICMPv6-ND: Sending NA for
FE80::21A:C4FF:FE29:1155 on GigE3/1
!The ER sends an NA to the GWR.
*Feb 15 21:35:30.302: ICMPv6-ND: Sending RA to FF02::1
on GigE3/1
*Feb 15 21:35:30.302: ICMPv6-ND: MTU=1500
```

Verifying IPv6 Addresses Assigned to GWR   Besides the configurations and negotiations above, notice the IPv6 address assigned via DHCPv6 on interface Ethernet1 and Ethernet0 on GWR. Ethernet1 of GWR is the upstream interface facing toward the SP ER. Ethernet0 is the downstream interface facing toward the host connected to the GWR. Note that GWR is acting as a DHCPv6 client for the ER and as a DHCPv6 server for the connected host.

```
GWR Router#show ipv6 interface ethernet1
Ethernet1 is up, line protocol is up
 IPv6 is enabled, link-local address is
FE80::207:EFF:FE03:6E65
 No Virtual link-local address(es):
 Global unicast address(es):
 2001:420:8:1:6:1:1:EBF1, subnet is
2001:420:8:1:6:1:1:EBF1/128 [CAL/PRE]
 ! Address assigned by DHCPv6.
 valid lifetime 1121384 preferred lifetime 516584

GWR Router#show ipv6 interface ethernet0
Ethernet0 is up, line protocol is up
 IPv6 is enabled, link-local address is
FE80::207:EFF:FE03:6E64
 No Virtual link-local address(es):
 Global unicast address(es):
 2001:420:8:1:7::1, subnet is 2001:420:8:1::/64 [CAL/PRE]
```

! Address assigned using DHCP-PD.
valid lifetime 1121385 preferred lifetime 516585

6.1.3.4   *Configuring SP ER as a DHCPv6-PD Server Using DUID*   This is the
same approach as that discussed in Section 6.1.3.3, where the ER also acts as a
DHCPv6-PD server in the absence of a central DHCPv6 server. In this model a
specific IPv6 prefix is assigned to a GWR by using DUID. Here, the
assumption is that the SP has visibility to each DUID tied to a GWR, which
makes this approach less attractive and scalable. For example, a cable MSO
that provisions millions of cable modems and set-top boxes would require
millions of DUIDs known and configured on the ER. If the GWR is replaced; a
new GWR would require careful reconfiguration of the same DUIDs on the
new GWR.

As ER is acting as a DHCPv6-PD server, all messages coming from GWR
are processed at the ER and responses sent to the GWR. The ER needs to be
preconfigured with specific DUID information in order to decide which IPv6
prefix should be delegated to the GWR. It also needs to keep track of the
list of prefixes delegated. In the example below, the ER is configured with
two DUIDs, where each DUID is assigned a different prefix. Interface Gigabit
Ethernet 1/0 is connected to a GWR and a DHCPv6 pool "cable" is tied to this
interface. (see Figure 6.7).

```
!
ipv6 cef
ipv6 unicast-routing
ipv6 dhcp pool cable
! DHCPv6 pool called ''cable.''
prefix-delegation 2001:1::/48 0003000100070E02BC97
! First DUID and associated IPv6 prefix.
prefix-delegation 2001:2::/48 0003000100070E02C873
! Second DUID and associated IPv6 prefix.
dns-server 2001:420:C0A8:C5E::1
domain-name rcdn-cable.cisco.com
!
interface Gigabit Ethernet 1/0
no ip address
ipv6 address 3FFE:B00:C18:2::1/64
ipv6 dhcp server cable
```

The following two important show commands are executed on the ER. They
show the DHCPv6 pool and bindings. In the **show ipv6 dhcp binding** output, one
DUID is listed for the only one active client, as displayed in the last line of the
show ipv6 dhcp pool command output. (see Figure 6.7).

```
ER_Router#show ipv6 dhcp binding
Client: FE80::4646:4628 (GigEthernet1/0)
```

```
DUID: 0003000100070E02BC97
IA PD: IA ID 0 × 00070001, T1 302400, T2 483840
  Prefix: 2001:1::/48
    preferred lifetime 604800, valid lifetime 2592000
    expires at Apr 25 2009 12:36 PM (2577505 seconds)
ER_Router#show ipv6 dhcp pool
DHCPv6 pool: cable
 Static bindings:
  Binding for client 0003000100070E02BC97
    IA PD: IA ID not specified; being used by 00070001
      Prefix: 2001:1::/48
        preferred lifetime 604800, valid lifetime 2592000
  Binding for client 0003000100070E02C873
    IA PD: IA ID not specified
      Prefix: 2001:2::/48
        preferred lifetime 604800, valid lifetime 2592000
DNS server: 2001:420:C0A8:C5E::1
Domain name: rcdn-cable.cisco.com
Active Clients: 1
```

The debug outputs are not shown for the case where DUID is used, but they are similar to those in previous sections where the ER router also acts as a DHCPv6-PD server. The following four messages are seen on an ER router performing the DHCPv6-PD server functionalities discussed in Section 6.1.3.3:

1. SOLICIT
2. ADVERTISE
3. REQUEST
4. REPLY

6.1.3.5 Configuring Common PPP-Based Models PPP-based deployment models and architectures discussed in Chapter 4 include two options that offer large-scale Internet access. This section includes some configuration examples on SP-managed routers for common PPP-based models. The GWR and AAA RADIUS server configurations for these models are shown in Chapters 5 and 7, respectively. Three PPP-based architecture models that offer large-scale Internet access are the ISP-operated deployment model, the wholesale deployment model, and the hybrid deployment model.

These models are based on PPP/RADIUS and do not necessarily include a dedicated DHCPv6 server. If a DHCPv6 server is needed, its functions are fulfilled by the router terminating the PPP sessions. This approach is commonly used to offer IPv4 Internet services, and many SPs adopted the model for offering IPv6 connectivity.

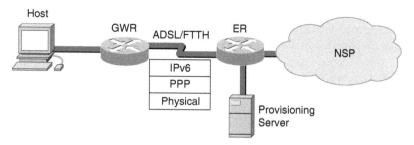

FIGURE 6.8 ISP-operated model.

ISP-Operated Deployment Model Configurations Figure 6.8 depicts the ISP-operated broadband access architecture with an SP ER. The RADIUS dialog between the ER and the AAA server occurs over an IPv4 transport.

The SP ER is connected to the GWR and there is no use of L2TP. User authentication is provided by a RADIUS server. ER-GWR link addresses and the prefixes delegated to the GWR are stored on the RADIUS server. The ER-GWR link does not need to be numbered with IPv6 Global addresses; IPv6 link-local addresses can be used. By using the **no ipv6 nd prefix framed-ipv6-prefix** command under **interface Virtual-Template1**, the prefix received from RADIUS will not be inserted in the RA and, instead, will be included in the DHCPv6 PD messages. A sample configuration on the ER for the ISP-operated model is as follows:

```
ipv6 unicast-routing
!
ip cef
ipv6 cef
!
aaa new-model
aaa authentication ppp default group radius
aaa authorization network default group radius
aaa authorization configuration PRELIST group radius
!
interface ATM0/0/0.1 point-to-point
 no ip directed-broadcast
 pvc 0/10
  encapsulation aal5mux ppp Virtual-Template1
  !
interface Virtual-Template1
 ipv6 enable
 no ipv6 nd suppress-ra
no ipv6 nd prefix framed-ipv6-prefix
 ipv6 dhcp server RCDN-1
 ppp authentication chap
```

```
!
ipv6 dhcp pool RCDN-1
 prefix-delegation aaa method-list PRELIST
 !
radius-server host 192.168.2.48
radius-server key radius-password
```

Wholesale Deployment Model Configurations Figure 6.9 depicts the wholesale deployment model featuring LAC and LNS elements. The GWR-NAP link supports IPv6 in order to provide IPv6 services to the enduser. L2TP, which is used as the tunneling mechanism between the LAC and the LNS is operated over IPv4. The RADIUS dialog between the LNS and the AAA server is done over an IPv4 transport as well.

Below is the standard LAC configuration, which is not IPv6 specific.

```
aaa new-model
aaa authentication ppp default group radius
aaa authorization network default group radius
aaa accounting network default wait-start group radius!
vpdn-group 1
 request-dialin
  protocol l2tp
  domain domain.net
  initiate-to ip 192.168.10.33
  local name lac
 !
interface ATM0/0/0.1 point-to-point
pvc 0/10
  encapsulation aal5mux ppp Virtual-Template1
 !
interface Virtual-Template1
 ppp authentication chap
 !
radius-server host 192.168.2.48
radius-server key radius-password
```

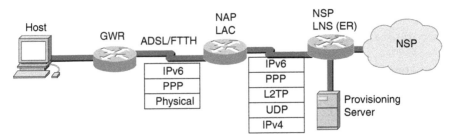

FIGURE 6.9 Wholesale deployment model.

The LNS configuration is shown below. The LNS delegates shorter than /64 prefixes, terminates the L2TP tunnels, and offers IPv6 connectivity to the enduser. The IPv6 prefixes are stored on the RADIUS server. When the ER-GWR link is not numbered with global addresses, the **no ipv6 nd prefix framed-ipv6-prefix** command is configured under **interface Virtual-Template1**. (see Figure 6.9).

```
ipv6 unicast-routing
!
ip cef
ipv6 cef
!
aaa new-model
aaa authentication ppp default group radius
aaa authorization network default group radius
aaa authorization configuration PRELIST group radius
!
vpdn enable
!
vpdn-group 1
 accept-dialin
  protocol l2tp
  virtual-template 1
 terminate-from hostname lac
 local name lns
!
interface Virtual-Template1
 ipv6 enable
 no ipv6 nd suppress-ra
 no ipv6 nd prefix framed-ipv6-prefix
 ipv6 dhcp server RCDN-1
 ppp authentication chap callin
 !
ipv6 dhcp pool SRV-P1
 prefix-delegation aaa method-list PRELIST
radius-server host 192.168.2.48
radius-server key radius-password
```

Hybrid Deployment Model Configurations The Cisco IOS routed bridged encapsulation (RBE) feature is used in the hybrid deployment approach as discussed in depth in Chapter 4. This approach is also known as the bridged access model. The following BRAS configuration applies to DSL access with RBE encapsulation. This configuration will allow Layer2 packets to be terminated on the BRAS and Layer3 (IPv6) packets to be routed to the ER. Note that unlike the configuration for IPv4 RBE, the ATM interface is numbered.

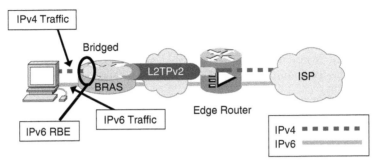

FIGURE 6.10 Hybrid deployment model.

Following is the BRAS configuration enabled for RBE, as shown in Figure 6.10.

```
ipv6 unicast-routing
!
ip cef
ipv6 cef
!
interface ATM0/0/0.1 point-to-point
 atm route-bridged ipv6
 ipv6 address 2001:db8:6::1/64
 pvc 0/10
  encapsulation aal5snap
```

6.2 SUMMARY

There are several ways of deploying IPv6 in SP core networks. Depending on the IPv4 deployment model, the existing IPv4 network infrastructure can be leveraged for IPv6 transport as well. The SP may choose to deploy IPv6 natively using the dual-stack approach or tunnel IPv6 traffic over its core network simply by upgrading the ERs (the tunnel endpoints) to dual-stack mode without having to upgrade all the intermediate devices. The tunneling approach can be a very cost-effective way for the SP to introduce IPv6 into the existing network. Some commonly used tunneling mechanisms were covered in this chapter. If the SP has an IP-based core, the dual-stack or IPv6 in IPv4 tunneling may be a good option for IPv6. If the SP network is based on an MPLS core, using the MPLS 6PE or 6VPE approach would be a more suitable way to easily deploy IPv6 in the network.

In this chapter we covered configuration scenarios and relevant show command outputs in illustrating IPv6 deployment and provisioning options on the ER. The configuration options for the GWR are presented in Chapter 5, and provisioning of services such as DHCPv6, TFTP, DNS, and AAA is covered in Chapter 7.

REFERENCES

1. S. Asadullah and A. Ahmed, "IPv6 in Broadband," Cisco Systems, Inc. Packet Magazine, Fourth Quarter 2004.
2. S. Asadullah, A. Ahmed, C. Popoviciu, P. Savola, and J. Palet, "ISP IPv6 Deployment Scenarios in Broadband Access Networks," RFC4779, January 2007.
3. B. Lourdelet, "Application Note: IPv6 Access Services," Cisco Systems, Inc.
4. B. Lourdelet, "Application Note: DHCPv6," Cisco Systems, Inc.
5. Cisco Systems, Inc., tutorial on IPv6 basics: "The ABC of IPv6."
6. S. McFarland, "RST-3217: IPv6 Advanced Concepts," Cisco Networkers 2004, January 2004.
7. S. Asadullah, and A. Ahmed, "BRKIPM-3300: Service Provider IPv6 Deployment," Cisco Networkers 2007, January 2007.
8. B. Aboba, G. Zorn, and D. Mitton, "RADIUS and IPv6," RFC3162, August 2001.
9. R. Droms et al. "Dynamic Host Configuration Protocol for IPv6 (DHCPv6)," RFC3315, July 2003.

7 Configuring and Troubleshooting IPv6 on Provisioning Servers

For service providers (SPs) to deploy IPv6 successfully on a large scale, it is critical for them to upgrade their back-end servers to support IPv6. Being able to provision and manage a large number of devices with IPv6 is a key factor in IPv6 adoption in SP environments, along with IPv6-based applications and value-added services for end users. In this chapter we discuss how to configure and troubleshoot various back-end servers and such applications as DHCP, DNS, and TFTP for IPv6 operation. Since it is difficult to cover several different implementations of each of these applications, we have chosen a product by a well-known vendor for illustration purposes: the Cisco network registrar (CNR).

7.1 IPv6 SUPPORT ON DHCP SERVERS

Due to the huge IPv6 address space and enormous number of IPv6 addresses available, it is critical for an SP to be able to assign and manage IPv6 addresses dynamically for a large number of devices in the network. In IPv4, DHCPv4 is commonly used for dynamic address assignment of end devices. Similarly, in IPv6, DHCPv6 can be used to assign and manage IPv6 addresses dynamically for end devices. As described in Chapter 2, DHCPv6 is similar to DHCPv4 in many ways and uses a similar four-way handshake between the client and the server. One of the key differences is that DHCPv4 relies on broadcast for sending certain messages between the client and server, whereas IPv6 relies on multicast to provide the same functionality. Additionally, there are certain DHCP options that have been added to DHCPv6 to support IPv6. These options are listed in Appendix B.

7.1.1 DHCPv6 Support in a Cisco Network Registrar

The Cisco network registrar (CNR) is a software product that provides DHCP, DNS, and TFTP server functionality to clients. CNR is supported on multiple

Deploying IPv6 in Broadband Access Networks, By Adeel Ahmed and Salman Asadullah
Copyright © 2009 John Wiley & Sons, Inc.

operating systems, such as Windows XP, Windows Server 2003, and certain Solaris and Linux versions. Starting with software release 6.2, CNR supports DHCPv6 and IPv6 DNS records but not TFTPv6. For Windows, the minimum release to support DHCPv6 is Windows XP with SP2. The DHCPv6 service requires that the server operating system (OS) supports IPv6 and that at least one interface on the server be configured for IPv6.

CNR release 7.0 introduces support for more advanced IPv6 features, such as DHCPv6 Leasequery (RFC5007), DNS updates for DHCPv6 (RFC4701, RFC4703, and RFC4704); and DHCPv6 CNR extensions. The CNR 7.0 software is supported on Windows Server 2003, Solaris 9 or 10, and Red Hat Enterprise Linux ES 4.0.

Some of the key DHCPv6 features supported in CNR include the following:

- *Stateless autoconfiguration* (also known as SLAAC and sometimes called DHCPv6lite per RFC 3736). The DHCPv6 server does not assign addresses but provides configuration parameters (such as DNS server information) to these clients.
- *Stateful configuration* (full DHCPv6 per RFC 3315). The DHCPv6 server assigns (permanent or temporary) addresses and provides configuration parameters to clients.
- *Links and prefixes* Links and prefixes define the network topology; Similar to DHCPv4 networks and scopes. Each link can have one or more prefixes.
- *Prefix delegation* (per RFC 3633). The DHCPv6 server delegates prefixes to clients (routers).
- *Client classes* Clients can be classified and prefixes selected based on known clients or packet-based expressions.
- *Policies and options* Attributes and options can be assigned to links, prefixes, and clients.
- *Static reservations* Clients can receive predetermined addresses.
- *Lease affinity* Lease affinity allows the server to give the same address to a client even if that client does not renew before the valid lifetime expires.
- Provisioning of virtual private networks (VPNs).
- *Statistics collection and logging* These monitor server activities.

7.1.2 Configuring DHCPv6 on CNR

In this section we discuss how DHCPv6 can be configured on a CNR server. Currently, there are two user interfaces for administering CNR: a web-based user interface (Web UI) and a command line interface (CLI):

- The Web UI runs in a browser window. The minimum version requirements are Microsoft Internet Explorer 6.0 (Service Pack 2), Mozilla

FIGURE 7.1 CNR login screen.

Firefox 1.5, or Netscape 7.0 and require installation of the Java runtime environment (JRE 5.0 [1.5]).

- The CLI runs in a Windows, Solaris, or Linux command window.

It will be easier to use the Web UI for basic CNR configuration. To access the CNR server using the Web UI, the user can point a web browser to the IP address of the CNR server and port 8080: for example, http://10.89.240.254:8080. The user should see the CNR login page shown in Figure 7.1. The default login name is "admin" and the password is "changeme." The username and password can be changed after logging into the server.

The screen that is displayed, after logging into the CNR server is shown in Figure 7.2.

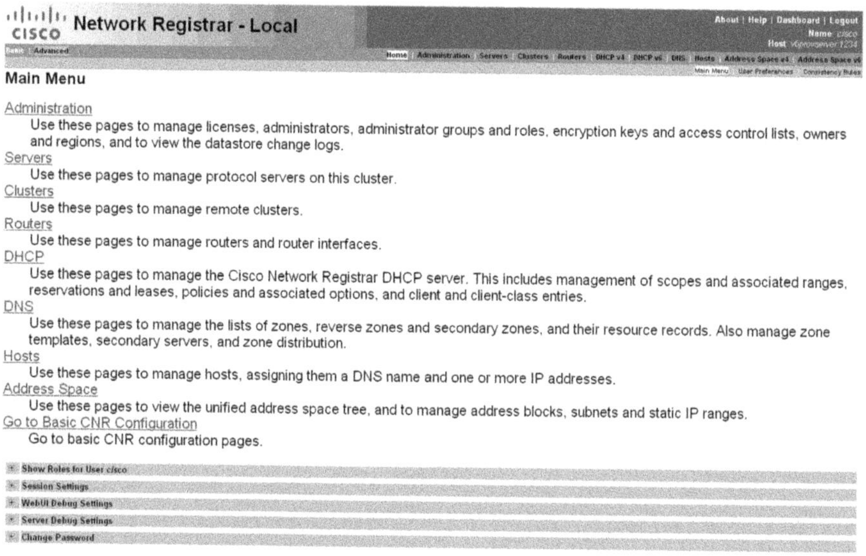

FIGURE 7.2 CNR main menu.

7.1.2.1 Configuring IPv6 Options and Policies in CNR   The first step in configuring the DHCPv6 server is to define the options and policies used when assigning IPv6 addresses and other network parameters to clients. A *policy* is a set of options that allow you to group lease times and other configuration parameters that a DHCPv6 server assigns to a client. These parameters are called *DHCPv6 options*. To configure DHCPv6 options, click on the **DHCPv6→Options** link under the advanced menu. This should bring up the DHCPv6 options definitions sets screen. Under the DHCPv6 option definition sets, click on the **dhcp6-config** link, then the **Modify Option Definition Set**, to see a list of DHCPv6 options available, shown in Figure 7.3.

These options are used when defining DHCPv6 policies in CNR. New options can be added, if not displayed in the list of available options, by clicking on the **Add Option Definition** button. Once the desired DHCPv6 options are available, the next step is to define policies that will be used by the server when replying to requests from clients.

To define the policies, click on the **Policies** link under DHCPv6 tab. This will display a list of current policies that exist on the server. A newly installed CNR server should only have the **default** and **system_default_policy** listed. New policies can be added by clicking on the **Add Policy** button. When adding a new policy, a list of available options is displayed in the pull-down menu. Figure 7.4 provides an example of a user-defined policy for IPv6.

List of Option Definitions for *dhcp6-config*		
Number	**Name**	**Type**
1	client-identifier	binary
2	server-identifier	binary
⊟3	ia-na	binary
0	iaid	unsigned 32-bit
0	t1	unsigned time
0	t2	unsigned time
⊟4	ia-ta	binary
⊟5	iaaddr	binary
0	address	IPv6 address
0	preferred-lifetime	unsigned time
0	valid-lifetime	unsigned time
6	oro	unsigned 16-bit
7	preference	unsigned 8-bit
8	elapsed-time	unsigned 16-bit
9	relay-message	binary
⊟11	auth	binary
12	server-unicast	IPv6 address
⊟13	status-code	binary
14	rapid-commit	zero size
⊟15	user-class	counted-type
16	vendor-class	vendor-class
⊟17	vendor-opts	vendor-opts
18	interface-id	binary
⊟19	reconfigure-message	unsigned 8-bit
20	reconfigure-accept	zero size
⊟21	sip-servers-name	DNS name
22	sip-servers-address	IPv6 address
⊟23	dns-servers	IPv6 address
24	domain-list	DNS name
⊟25	ia-pd	binary
0	iaid	unsigned 32-bit
0	t1	unsigned time
0	t2	unsigned time
⊟26	iaprefix	binary
0	preferred-lifetime	unsigned time
0	valid-lifetime	unsigned time
0	prefix-length	unsigned 8-bit
0	prefix	IPv6 address

FIGURE 7.3   List of available DHCPv6 options.

View DHCP Policy *CM_IPv6_Policy*

Attribute		Value		
Name		CM_IPv6_Policy		
Offer Timeout		2m		
Grace Period		5m		
⊟ DHCPv4 Options	Name		Number		Legacy	Value
⊟ DHCPv6 Options				Value		
Configured Options	[23] (dhcp6-config) dns-servers		(IPv6 address)	2001:bb:cab:10:203:baff:fe1f:bc68	
	[24] (dhcp6-config) domain-list		(DNS name)	rcdn-cable.cisco.com:	
	[12] (dhcp6-config) server-unicast		(IPv6 address)	2001:bb:cab:10:203:baff:fe1f:bc68	
	Name	Number			Value
	Name	Number		Value	
Configured Options	[17] (dhcp6-cablelabs-config)	cablelabs-17	(vendor-opts)	(enterprise-id 4491 ((tftp-servers 32 2001:bb:cab:10:203:baff:fe1f:bc68)(config-file-name 33 "docsis11_generic_WB.cm")))	

FIGURE 7.4   User-defined policy for IPv6.

This example shows options 23 (dns-server), 24 (domain-list) and 12 (server-unicast) defined for the policy **CM_IPv6_Policy**. Also listed are CableLabs vendor-specific options such as option 32 (tftp-servers) and option 33 (config-file-name). When a client sends a DHCPv6 SOLICIT message, the server can respond back with a DHCPv6 ADVERTISE message and include in its response the options listed above, based on the configured policy.

Besides providing clients with the DHCPv6 options requested, the server may need to assign the clients IPv6 addresses, or delegate IPv6 prefixes in the case of DHCP-PD. In the following sections we describe how the CNR server can be configured for IPv6 address management.

7.1.2.2 Configuring IPv6 Links and Prefixes in CNR   To configure the CNR DHCPv6 server to assign IPv6 addresses and prefixes to clients, the user needs to define criteria that help the server decide which prefix to use for address assignment. These criteria can be defined by configuring **Links** and **Prefixes** in CNR.

- *Link* A network segment that can have one or more associated prefixes. It provides an additional layer at which policies can be defined for DHCPv6 clients. An IPv6 prefix's link attribute is similar to an IPv4 scope's primary scope, except that the link attribute names a link and not another prefix. A link needs to be created only if more than one prefix object with a different IPv6 prefix exists on a link. When the server loads the configuration, if a prefix has no explicit link, the server searches for or creates an implicit link.

- *Prefix* Equivalent to a scope or subnet in IPv4. The link associated with an IPv6 prefix is similar to an IPv4 scope's primary scope except that it names a link and not another prefix. Just as with scopes, multiple prefix objects for the same IPv6 prefix may be created. However, rather than supporting multiple ranges with explicit start and end addresses, prefixes support only a single range that must be an IPv6 prefix with a length the

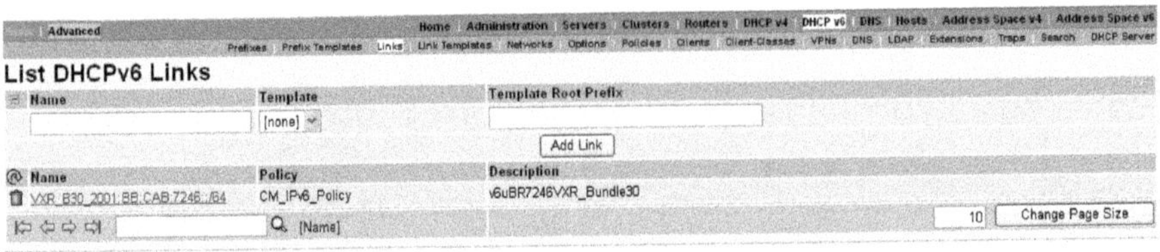

FIGURE 7.5   Configuring DHCPv6 links in CNR.

same as, or longer than, that of the prefix. If a range is not specified, the IPv6 prefix is used as the range. A range is limited to powers of 2. The range must be unique (it cannot be duplicated by any other range, except in a different VPN) and cannot be contained in, or contain, another range.

To configure a link, click on the **Links** option under the DHCPv6 menu. Configure the name of the link and click on the **Add Link** button. Any previously configured links will be listed on this screen, as shown in Figure 7.5.

To view the detailed configuration of links, the user clicks on the link name, which will display the link configuration shown in Figure 7.6.

Besides configuring the name and description of the link, the user can also assign a specific policy and associate specific prefixes to the link. Once the link is configured, the next step is to configure the prefixes that will be associated with this link. To configure IPv6 prefixes in CNR, click on the **Prefixes** option under the DHCPv6 menu. This will display the IPv6 prefixes currently configured and list the option to configure new IPv6 prefixes, as illustrated in Figure 7.7. To configure the new prefix, define the name of the prefix and configure the address range and the length of the prefix. The default prefix length is /64. The **DHCP Type** pull-down menu allows the user to define the type of prefix being

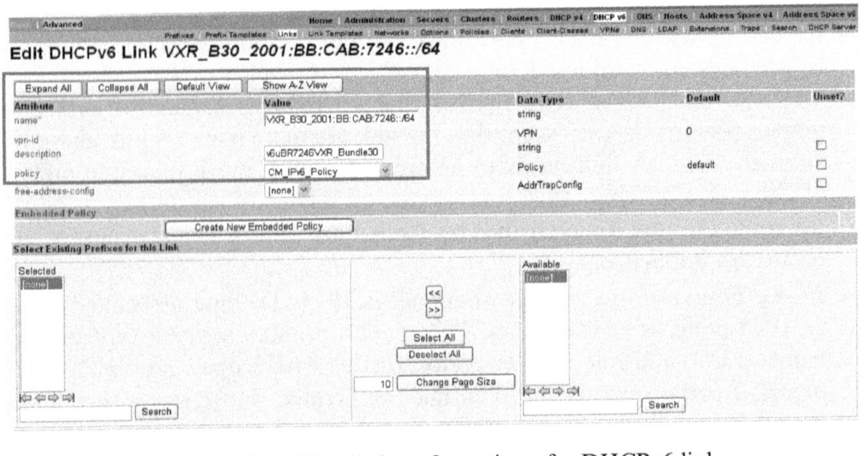

FIGURE 7.6   Detailed configuration of a DHCPv6 link.

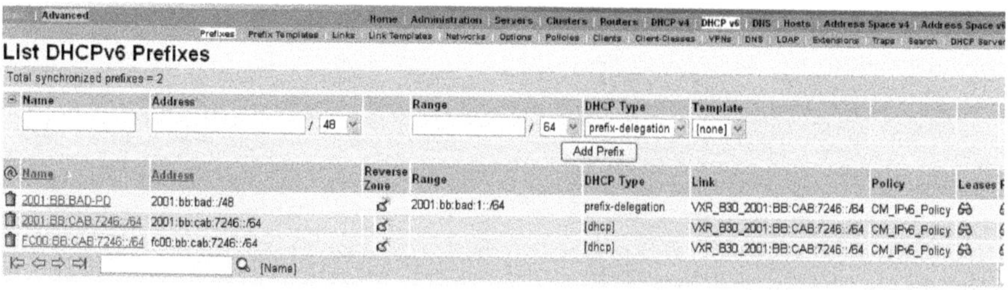

List DHCPv6 Prefixes

Total synchronized prefixes = 2

Name	Address		Range		DHCP Type	Template	
		/ 48 ∨		/ 64 ∨	prefix-delegation ∨	[none] ∨	
				Add Prefix			

Name	Address	Reverse Zone	Range	DHCP Type	Link	Policy	Leases
2001-BB-BAD-PD	2001:bb:bad::/48		2001:bb:bad:1::/64	prefix-delegation	VXR_B30_2001:BB:CAB:7246::/64	CM_IP6_Policy	
2001:BB:CAB:7246::/64	2001:bb:cab:7246::/64			[dhcp]	VXR_B30_2001:BB:CAB:7246::/64	CM_IP6_Policy	
FC00:BB:CAB:7246::/64	fc00:bb:cab:7246::/64			[dhcp]	VXR_B30_2001:BB:CAB:7246::/64	CM_IP6_Policy	

FIGURE 7.7   Configuring IPv6 prefixes in CNR.

configured and whether the prefix is going to be used for stateful DHCP (dhcp), SLAAC (stateless), or DHCP-PD (prefix delegation). Click on the name of the prefix to view the detailed prefix configuration, shown in Figure 7.8.

The figure displays the details of the 2001:BB:CAB:7246::/64 prefix. The link associated with this is VXR_B30_2001:BB:CAB:7246/64 and the prefix type is **dhcp.** Therefore, this prefix will be used for stateful DHCP to assign clients an IPv6 address as well as other network parameters through the policy and options configured for this prefix. Since the prefix will be used for stateful DHCP, the assumption is that there is probably a router in the network acting as a DHCPv6 relay in order to forward DHCPv6 packets between the client and the DHCPv6 server. The corresponding interface level configuration for a router acting as the DHCPv6 relay is as follows:

```
interface Bundle30
! Name of the interface being configured.
no ip address
! This is an IPv6-only interface, so no IPv4 address is
configured.
ipv6 address 2001:BB:CAB:7246::2/64
! Global unicast IPv6 prefix for this interface.
ipv6 address FC00:BB:CAB:7246::2/64
! Unique local IPv6 prefix for this interface.
ipv6 nd managed-config-flag
! This informs the clients that stateful DHCP can be used
for IPv6 address assignment.
ipv6 nd other-config-flag
! This informs clients that other network parameters can be
requested from the DHCPv6 server.
ipv6 dhcp relay destination
2001:BB:CAB:10:203:BAFF:FE1F:BC68
! This tells the router acting as a DHCPv6 relay where to
forward DHCPv6 messages to. This is the IPv6 address of the
DHCPv6 server.
```

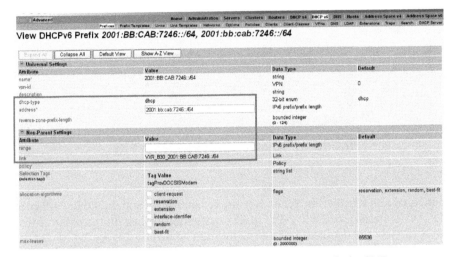

FIGURE 7.8 Detailed configuration of an IPv6 prefix in CNR.

Similarly, you can assign another IPv6 prefix for DHCP-PD for GWR that have multiple networks connected to them, and they act as DHCP-PD requesting routers. When the server receives a request for prefix delegation it will assign a prefix instead of assigning a single IPv6 address to the requesting client. The DHCP-PD configuration is illustrated in Figure 7.9. The IPv6 prefix configured for DHCP-PD is 2001:BB:BAD::/48, the link associated with this

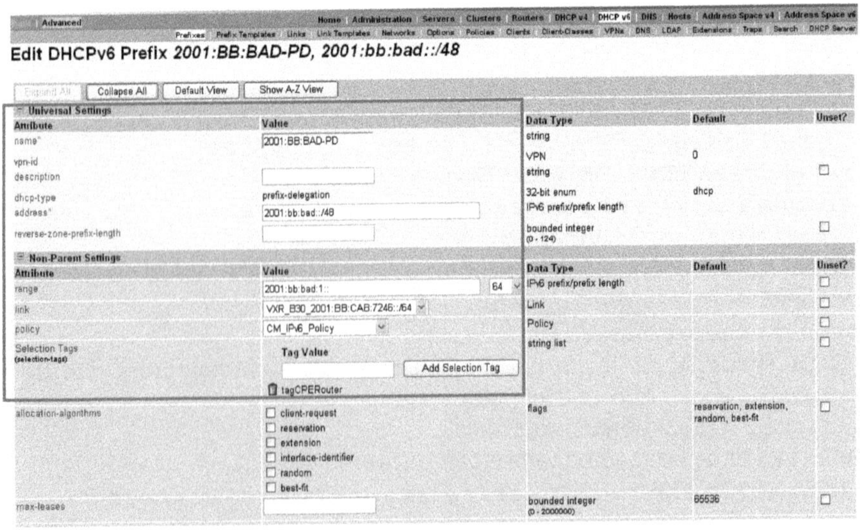

FIGURE 7.9 DHCP-PD configuration in CNR.

prefix is VXR_B30_2001:BB:CAB:7246::/64, and the policy is CM_IPv6_ Policy. The address range starts at 2001:BB:BAD:1::/64. Also note that the dhcp-type is **prefix-delegation**.

Determining Links and Prefixes on the DHCPv6 Server in CNR DHCPv6 clients are identified by their DHCP unique identifier (DUID), which is the DHCPv4 concept of hardware addresses and client IDs consolidated into one unique client identifier. When the DHCPv6 server receives a DHCPv6 message from a client, it determines the link and prefixes to be used to service the request as follows:

1. Finds the source address:
   a. If the client message was relayed, the server sets the source address to the first nonzero link-address field, starting with the relay-forward message closest to the client. If the server finds a source address, it proceeds to step 2.
   b. Otherwise, if the message source address is a link-local address, the server sets the source address to the first address for the interface on which the message was received and for which a prefix exists (or 0 if no prefix is found for any address). It then proceeds to step 2.
   c. Otherwise, the server sets the source address to the message source address.
2. Locates the prefix for the source address. If the server cannot find a prefix for the source address, it cannot service the client and drops the request.
3. Locates the link for the prefix. This always exists and is either an explicitly configured link or an implicitly created link based on the prefix address.

Now that the server can determine the client's link, it can process the client's request. Depending on whether the client's request is stateful DHCP or prefix-delegated along with selection criteria and other factors, one or more prefixes for the link might be used to service the client's request. This is one area of difference between DHCPv4 and DHCPv6. In DHCPv4, only one of the scopes from the network is selected to service the client's request. In DHCPv6, all of the prefixes for the link are potentially usable. Thus, a client might be assigned an address, or delegated a prefix, from multiple prefixes for the link (subject to selection criteria and other conditions).

Generating IPv6 Addresses and Delegating IPv6 Prefixes IPv6 addresses are 128 bits; in most cases the DHCPv6 servers assign at least 64 of those bits. Because of this, addresses are generated using the client's 64-bit interface identifier whenever possible, or through a random number generator. The interface identifier emulates how SLAAC assigns addresses to the client. Unfortunately, there are privacy concerns regarding SLAAC, and it does not work if the client requests multiple addresses on the same prefix. To address this issue, the alternative is the use of random number generator (see RFC3041). The DHCPv6 server uses a first-fit algorithm when generating delegated prefixes. Depending on the configured default, minimum, and maximum, and the client's requested prefix length, it returns the first available shortest prefix to the client.

Configuring DHCPv6 Client Class and Clients in CNR   A client class can be configured if certain group of clients needs to be assigned an IPv6 address from a specific prefix or if the group of clients need specific DHCPv6 options. A client class is tied to the IPv6 prefix using certain fields, such as **selection-tags** under the IPv6 prefix and **selection-criteria** under the client class. To configure a client class in CNR, click on the **Client-Classes** option under the DHCPv6 menu. This will list any existing client classes configured and also list the option to add a new client class. To add a new client class click on the **Add Client-Class** button and fill out the appropriate fields, such as the name of the client class, the domain name, the policy name, the selection criteria and the v6 client lookup id, as illustrated in Figure 7.10. The **v6-client-lookup-id** specifies the key value used to lookup the DHCPv6 client in the client database by using an expression that evaluates to a string or blob that is a valid string.

In the figure, the name of the client class is **IPv6_CM**, the domain name is **rcdn-cable.cisco.com**, the policy associated with this client class is **CM_IPv6_ Policy**, the selection criteria field is set to **tagProvDOCSISModem**, and the v6 client lookup id is configured with a value of **00:03:00:01:00:18:68:35:27:b6**. One thing to note is that the string defined for the selection criteria needs to match exactly the string defined for the selection-tags field under the IPv6 prefix which is being associated with this client class.

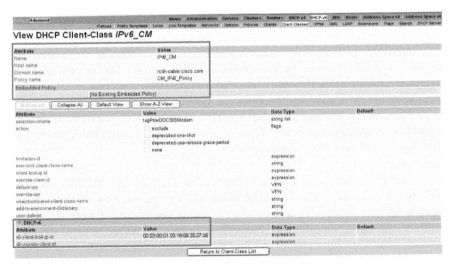

FIGURE 7.10 Configuring DHCPv6 client class in CNR.

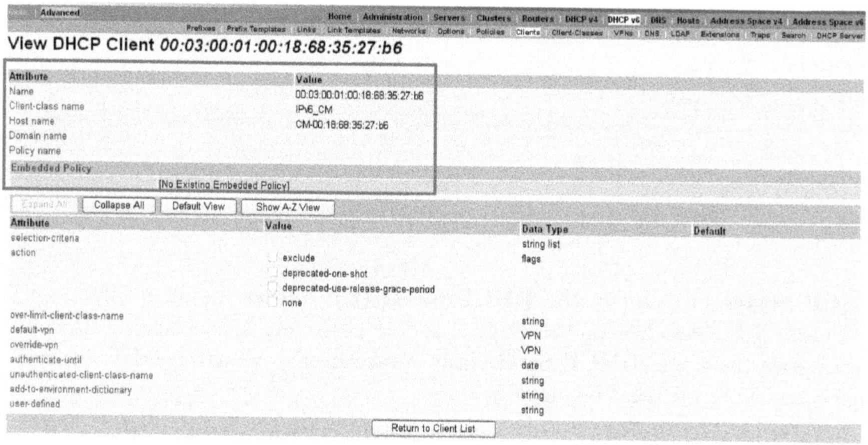

FIGURE 7.11   DHCPv6 client configuration details.

Once the client class is configured, multiple clients can be tied to the client class using the clients menu option under the DHCPv6 menu. Clients can be associated with a specific client class using the client-class name under the client configuration, as shown in Figure 7.11. Here the client is tied to a client-class name called **IPv6_CM**. The client is also configured with a host name that will be assigned to this client by the server when the client exchanges DHCPv6 messages with the server.

When the server is processing the request from clients, it will use the client class to match the request to the appropriate prefix using the criteria defined. An example of this is listed in the following CNR DHCP log:

```
Client Class lookup detail for Client DUID:
00:03:00:01:00:18:68:52:7b:2c from Request 'R3' Client
Class Name='IPv6_CM,' client-lookup-id result: ''Specified
host name: no, Host Name='none', Specified domain name: yes,
Domain Name='rcdn-cable.cisco.com', Client
Policy='none', Client Class Policy='CM_IPv6_Policy',
Action='none', InclusionCriteria='tagProvDOCSISModem',
ExclusionCriteria='', over-limit-client-class='', Client
Class Name Derived From: client-class-lookup-id.
```

The server does a client lookup based on the DUID. It finds a match for the DUID and uses the **IPv6_CM** client class to process this client's request.

7.1.3   Troubleshooting a CNR DHCPv6 Server

There are several levels of debugging options available in CNR to troubleshoot DHCPv6-related issues. To enable these debugging options, click on the **Local**

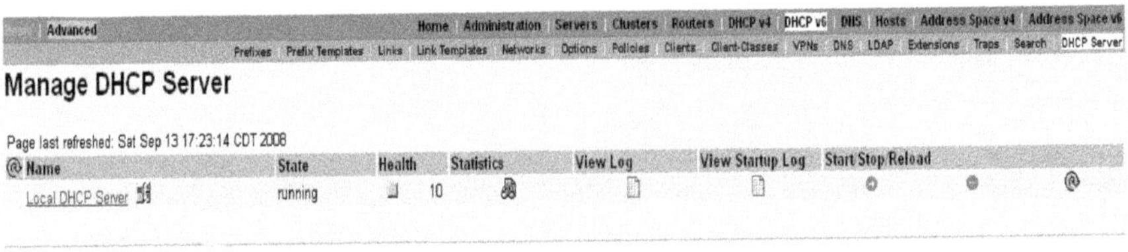

FIGURE 7.12   Editing the DHCP server parameters.

DHCP Server link under the **DHCPv6→DHCP Server** menu as illustrated in Figure 7.12. As shown, the DHCP server parameters can be modified by clicking on the **Local DHCP Server** link. You can also view the DHCP server logs by clicking on the **View Log** option.

Several logging options and levels are available on the CNR DHCP server, shown in Figure 7.13. Once the logging options are enabled, click on the **Modify Server** tab at the bottom of the screen to apply the changes and then reload the server by clicking on the reload icon. To view the logs, click on the **View Log** icon. The logs can also be viewed by using the UNIX tail command in a terminal window(UNIX) or cmd window(Windows). You use the **tail -f /var/nwreg2/local/logs/name_dhcp_1_log** command to view the details of the DHCPv6 messages on the DHCPv6 server. The server will display detailed logs as shown in Figure 7.14.

As you can see from the logs, the server is displaying the contents of a client's DHCPv6 SOLICIT message forwarded to the server by a DHCPv6 relay. The SOLICIT message contains the request for IPv6 address assignment using

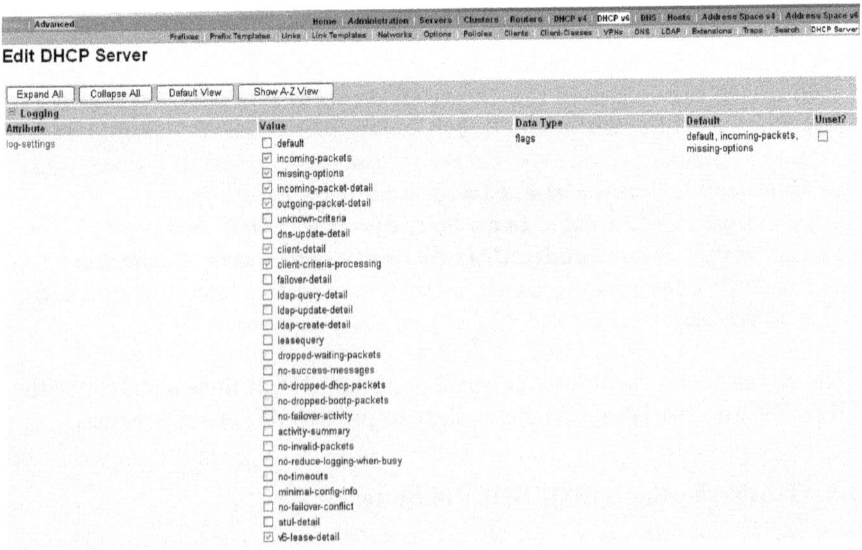

FIGURE 7.13   Enabling logging on CNR DHCP server.

```
----- RECEIVED -- R3 -----
R3:    ->  received 273 bytes from 2001:BB:CAB:30::1, port 547
R3:    ->  +- Start of RELAY-FORW (12) message (273 bytes)
R3:    ->  |   hop-count 0,
R3:    ->  |   link-address 2001:BB:CAB:7246::1,
R3:    ->  |   peer-address fe80::218:68ff:fe52:7b2c
R3:    ->  |   relay-message (9) option (187 bytes)
R3:    ->  |   +- Start of SOLICIT (1) message (187 bytes)
R3:    ->  |   |   transaction-id 12378582
R3:    ->  |   |   rapid-commit (14) option (0 bytes)
R3:    ->  |   |   vendor-opts (17) option (120 bytes)
R3:    ->  |   |     (enterprise-id 4491,
R3:    ->  |   |     ((oro 1 32,33,34,40,41),
R3:    ->  |   |     (modem-capabilities 35 05:56:01:01:01:02:01:02:03:01:01:04:01:01:05:01:01:06:01:01:07
R3:    ->  |   |     (device-id 36 00:18:68:52:7b:2c)))
R3:    ->  |   |   vendor-class (16) option (15 bytes)
R3:    ->  |   |     (enterprise-id 4491,
R3:    ->  |   |     ((00:09:64:6f:63:73:69:73:33:2e:30)))
R3:    ->  |   |   client-identifier (1) option (10 bytes)
R3:    ->  |   |     00:03:00:01:00:18:68:52:7b:2c
R3:    ->  |   |   ia-na (3) option (12 bytes)
R3:    ->  |   |     (iaid 1750235948, t1 0, t2 0)
R3:    ->  |   |   elapsed-time (8) option (2 bytes)
R3:    ->  |   |     11
R3:    ->  |   +- End of SOLICIT message
R3:    ->  |   interface-id (18) option (18 bytes)
R3:    ->  |     42:75:33:26:43:61:34:2f:30:00:00:18:68:52:7b:2c:00:00
R3:    ->  |   vendor-opts (17) option (22 bytes)
R3:    ->  |     (enterprise-id 4491,
R3:    ->  |     ((cmts-capabilities 1025 01:02:03:00),
R3:    ->  |     (cm-mac-address 1026 00:18:68:52:7b:2c)))
R3:    ->  +- End of RELAY-FORW message
R3:    ----- END OF RECEIVED -- R3 -----
```

FIGURE 7.14 DHCP server logs for DHCPv6 messages.

option 3 (IA-NA) and the options request option (**ORO**) numbers 32, 33, 34, 40, 41 [tftp-servers (32), config-file-name (33), syslog-servers (34)]. Once the server receives this request, it needs to check the configured policy to see if it can provide the client with and the options requested to if there is a matching IPv6 prefix configured from which it can assign an address to the client. If the server finds a match, it replies back with a DHCPv6 ADVERTISE message; otherwise, it drops the packet.

Troubleshooting DHCPv6 and DHCP-PD As explained in Chapter 2, DHCPv6 uses a four-way handshake between the client and the server. The client sends a DHCPv6 SOLICIT message to the server, the server responds with an ADVERTISE message, the client then sends a REQUEST, and finally, the server responds with a REPLY. If the client is unable to obtain an IPv6 address, it could be due either to the server dropping the client request or the packet getting dropped by an intermediate device between the client and the server. You can verify the DHCPv6 message exchange on the server by viewing the contents of the DHCP logs via the GUI interface or CLI (**/var/nwreg2/ local/logs/name_dhcp_1_log**). The server will report a warning message when it drops a DHCPv6 request from the client. An example is

08/21/2008 19:23:22 name/dhcp/1 Warning Protocol 0 18166 Received SOLICIT packet but found no Prefixes for source address '2001:72:1::2/64.' Dropping packet.

Since a matching prefix was not found, the server dropped the SOLICIT from the client. In this case the client would not be able to receive an IPv6 address from the server. If the DHCPv6 server is configured properly and receives a valid DHCPv6 SOLICIT message from the client, it will respond with an ADVERTISE, as illustrated in Figure 7.15.

In this example you can see that the server is offering a lease to the client, and the IPv6 address assigned to this client is 2001:1::2, with a preferred and valid lifetime of 24 hours.

When troubleshooting problems related to DHCP-PD, it is important to verify that the client has actually requested the server to delegate the prefix using option 25 (IA-PD). If everything is configured properly on the server, including the DHCP-PD prefix to be delegated with the appropriate length (shorter than a /64), the server should respond with a message similar to the one shown in Figure 7.16. A DHCP-PD prefix will be sent to the client using option 26 (IAPREFIX).

In the debugs shown, you can see that the server is assigning a 2001:420:8:: /48 prefix to the client for prefix delegation using option 26 (IAPREFIX). The server also includes the preferred (7 days) and valid (14 days) lifetimes associated with this prefix. Other network parameters included in the REPLY message are the IPv6 address using option 5 (IAADDR), DNS server address using option 23 (DNS-SERVER), and domain name using option 24 (DOMAIN-LIST).

7.2   IPv6 SUPPORT ON DNS SERVERS

The changes that need to be made to DNS to support IPv6 are defined in RFC3596. It describes the AAAA record type, the AAAA data format, the AAAA query, the textual format of AAAA records, and the format of the IPv6 address in the IP6.ARPA domain for reverse domain lookups. Another relevant IETF document that discusses representation of IPv6 addresses in DNS is RFC3363.

7.2.1   IPv6 Support on a DNS Server in a Cisco Network Registrar

As mentioned earlier in this chapter, starting with release 6.2, CNR supports IPv6 functionality on the DNS server. This includes full support of AAAA and PTR resource records for IPv6 addresses. In release 7.0, CNR introduces additional DNS support for IPv6, including AAAA and PTR mappings for DHCPv6 leases, and DNS updates for stateful IPv6 addresses, among other features.

7.2.2   Configuring a CNR DNS Server for IPv6

The DNS server needs to be configured with several options in order to support IPv4 and IPv6 functionality and for the server to respond to DNS queries from both types of clients. In this section we discuss the various options that need to be configured on a CNR DNS server.

```
Server received a relayed SOLICIT from Client: DUID: 00:03:00:01:00:18:68:52:7b:2c packet: R3 on network
interface ifindex 5, device 'eth0', 2 in use.
Lease 2001:32::14 OFFERED to Client: KEY: 00:03:00:01:00:18:68:52:7b:2c DUID: 00:03:00:01:00:18:68:52:7b:2c
packet: R3 with lifetimes 24h/24h.
X26: ----- TRANSMITTED -- X26 -----
X26: <-  transmitted 228 bytes to 2001:1::1, port 547
X26: <-  +- Start of RELAY-REPLY (13) message (228 bytes)
X26: <-      hop-count 0,
X26: <-      link-address  2001:32::1,
X26: <-      peer-address fe80::218:68ff:fe52:7b2c
X26: <-  interface-id (18) option (18 bytes)
X26: <-      42:75:33:26:43:61:34:2f:30:00:00:18:68:52:7b:2c:00:00
X26: <-  relaymessage (9) option (168 bytes)
X26: <-  +- Start of ADVERTISE (2) message (168 bytes)
X26: <-      transaction-id 12378582
X26: <-  client-identifier (1) option (10 bytes)
X26: <-      00:03:00:01:00:18:68:52:7b:2c
X26: <-  server-identifier (2) option (14 bytes)
X26: <-      00:01:00:01:44:3a:cd:25:00:03:47:9b:b9:f0
X26: <-  ia-na (3) option (40 bytes)
X26: <-      (iaid 1750235948, t1 12h, t2 19h12m)
X26: <-      iaaddr (5) option (24 bytes)
X26: <-      (address 2001:32::14,
X26: <-      preferred-lifetime 24h,
X26: <-      valid-lifetime 24h)
X26: <-  preference (7) option (1 bytes)
X26: <-      255
X26: <-  vendor-opts (17) option (61 bytes)
X26: <-      (enterprise-id 4491,
X26: <-      (tftp-servers 32 2001:1::2),
X26: <-      (config-file-name 33 wb.cm),
X26: <-      (rfc868-servers 37 2001:1::2),
X26: <-      (time-offset 38 5h)))
X26: <-  domain-list (24) option (14 bytes)
X26: <-      v6.cisco.com.
X26: <-  +- End of ADVERTISE message
X26: <-  +- End of RELAY-REPLY message
X26: ----- END OF TRANSMITTED -- X26 -----
```

FIGURE 7.15 DHCPv6 ADVERTISE message on the server.

171

```
*Feb 15 21:35:20.942: IPv6 DHCP: Sending REPLY to FE80::207:EFF:FE03:6E65 on GigE3/1
*Feb 15 21:35:20.942: IPv6 DHCP: detailed packet contents
*Feb 15 21:35:20.942:  src FE80::21A:C4FF:FE29:1155
*Feb 15 21:35:20.942:  dst FE80::207:EFF:FE03:6E65 (GigE3/1)
*Feb 15 21:35:20.942:  type REPLY(7), xid 13530568
*Feb 15 21:35:20.942:  option CLIENTID(1), len 10
*Feb 15 21:35:20.942:    0003000100070E036E65
*Feb 15 21:35:20.942:  option SERVERID(2), len 14
*Feb 15 21:35:20.942:    0001000143BF22B6080020E8FAC0
*Feb 15 21:35:20.942:  option IA-NA(3), len 40
*Feb 15 21:35:20.942:    IAID 0x00020001, T1 302400, T2 483840
*Feb 15 21:35:20.942:    option IAADDR(5), len 24
*Feb 15 21:35:20.942:      IPv6 address 2001:420:8:1:6:1:1:EBF1
*Feb 15 21:35:20.942:      preferred 604800, valid 1209600
*Feb 15 21:35:20.942:  option IA-PD(25), len 41
*Feb 15 21:35:20.942:    IAID 0x00020001, T1 302400, T2 483840
*Feb 15 21:35:20.942:    option IAPREFIX(26), len 25
*Feb 15 21:35:20.942:      preferred 604800, valid 1209600, prefix 2001:420:8::/48 [DHCP-PD]
*Feb 15 21:35:20.946:  option DNS-SERVERS(23), len 16
*Feb 15 21:35:20.946:    2001:420:3800:801:A00:20FF:FEE5:63E3
*Feb 15 21:35:20.946:  option DOMAIN-LIST(24), len 14
*Feb 15 21:35:20.946:    v6.cisco.com
```

FIGURE 7.16 DHCPv6 REPLY message for DHCP-PD.

7.2.2.1 *Configuring Forward and Reverse Zones in CNR* To configure a DNS server in CNR, click on the ***DNS*** tab under the ***Advanced*** menu option. This will bring up the DNS submenu and list the forward zones currently configured on the DNS server, as illustrated in Figure 7.17. This shows the forward zone **rcdn-cable.cisco.com** configured on the DNS server. To view the detailed configuration, click on the name of the zone and the next screen will display the various fields, such as start of authority (SOA) attributes, zone transfer settings, and DNS update settings. To add a new zone, enter the name of the zone and click on the **Add Zone** button. This will take you to the next screen, where you can enter the details about this zone. Once all the relevant information is entered, click on the **Modify Zone** button to save your changes.

To configure a reverse zone, click on the **Reverse Zones** sub menu option under the **DNS** menu. This will list all the existing reverse zones as well as the option to add new reverse zones, as in Figure 7.18, showing reverse zones currently configured on the DNS server. To view the detailed configuration of the reverse zone, click on the name of the zone. To view the host entries and resource records, click on the appropriate glasses icon in the **Hosts** or **RRs**

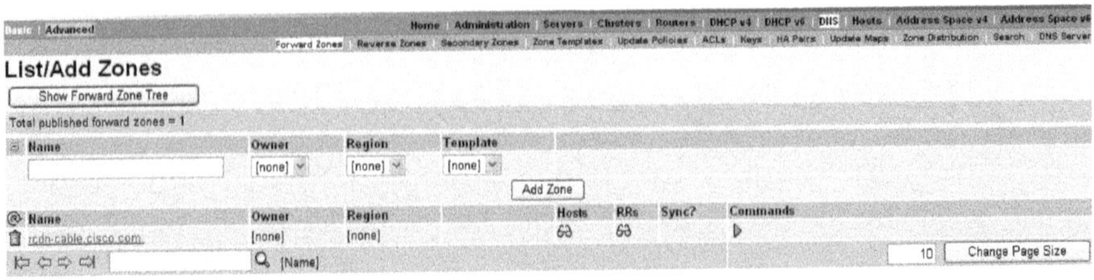

FIGURE 7.17 DNS submenu display.

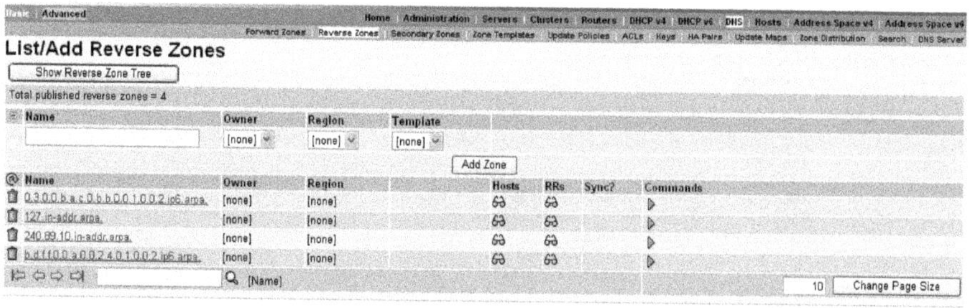

FIGURE 7.18   Adding and displaying reverse rones on the DNS server.

column. When configuring reverse IPv6 zones on the DNS server, using a multiple of 4 for the reverse-zone-prefix-length value is recommended because ip6.arpa zones are on 4-bit boundaries.

7.2.2.2   *Configuring Hosts and Resource Records in CNR*   To configure or view host or resource record information, click on the appropriate icon (Figure 7.18). You will be taken to the next screen, which displays the list of currently configured entries as well as listing the option to add new entries, as shown in Figure 7.19. This example displays several entries that currently exist on the DNS server. You can see that for dual-stack hosts, both the IPv4 (A record) and IPv6 (AAAA record) are listed with the corresponding addresses.

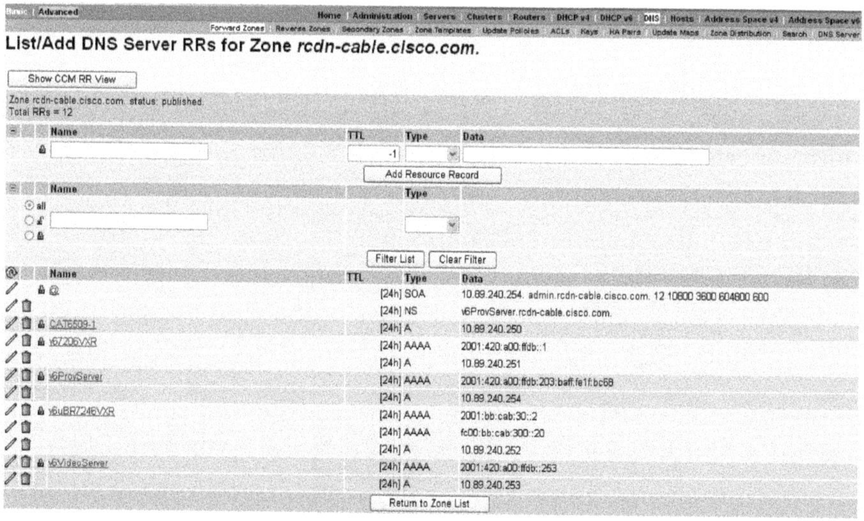

FIGURE 7.19   Configuring and adding RR on the DNS server.

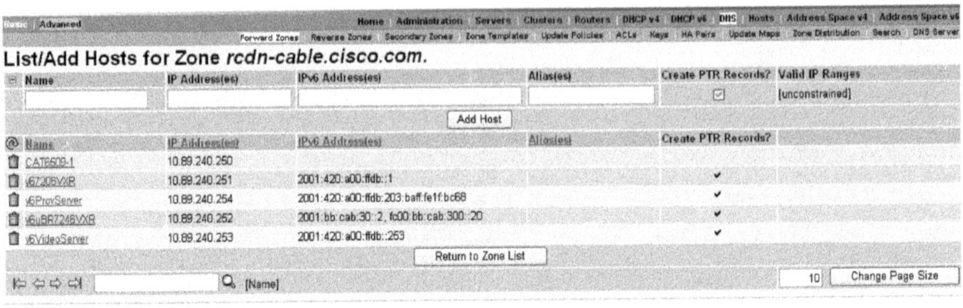

FIGURE 7.20 Displaying and adding hosts on the DNS server.

To add new or to list existing host records, click on the *glasses* icon in the **Hosts** column, which will take you to the next screen that displays host information (Figure 7.20). In this example we see several host entries listed with both IPv4 and IPv6 addresses. When creating new host entries, you can check the box **Create PTR Records?** to generate a reverse mapping for the IP address being configured. To view the configuration details of a particular host, click on the name of the host.

In Figure 7.21, you see the configuration details for the host **v6VideoServer**. This host has multiple IP addresses associated with it, including both IPv6 and IPv4 addresses. The host entry includes an A record as well as an AAAA record for forward lookups (hostname to IP address). To view the PTR record for this host for reverse lookups (IP address to hostname), click on the **resource record** icon next to the reverse zone under **DNS→Reverse Zones→Name**. An example of a resource record for the DNS reverse zone is shown in Figure 7.22.

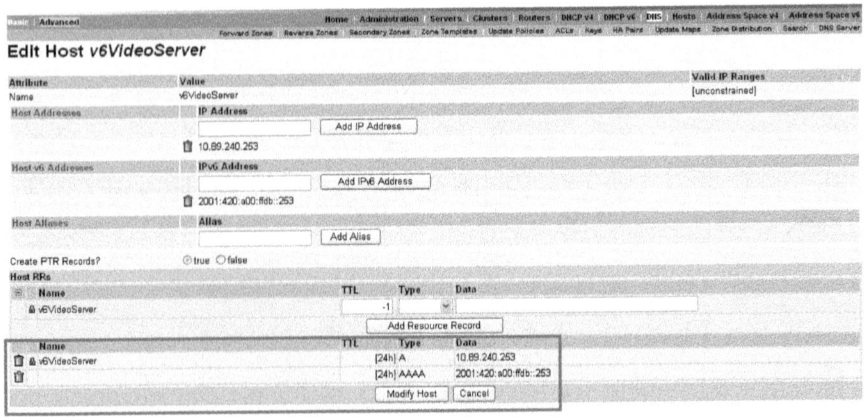

FIGURE 7.21 Displaying detailed configuration for a host entry on the DNS server.

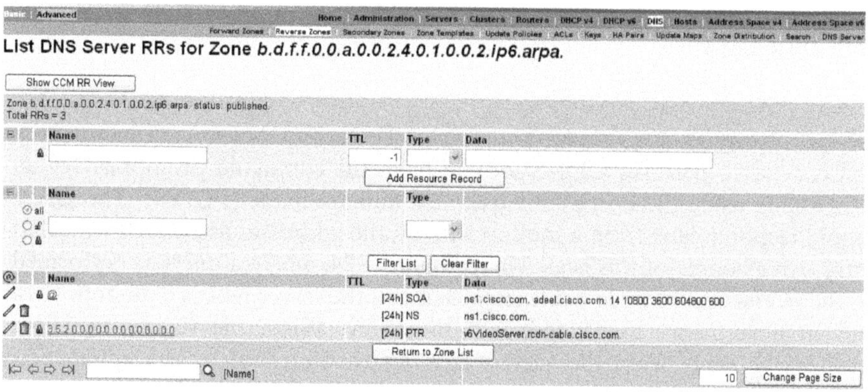

FIGURE 7.22   Resource record details for DNS reverse zone.

7.2.3   Troubleshooting a CNR DNS Server

Several debugging options are available in CNR to troubleshoot DNS-related problems. These options can be enabled via either the Web UI or CLI. To enable debugging using the Web UI, click on the **Local DNS Server** link under the **DNS→DNS Server** menu. The next screen will list the various options that can be enabled on the DNS server for debugging purposes, shown in Figure 7.23. To view the detailed logs, click on the log icon listed under the **View Log** column on the manage DNS server page. If the DNS server is configured properly with the correct forward zone information, including

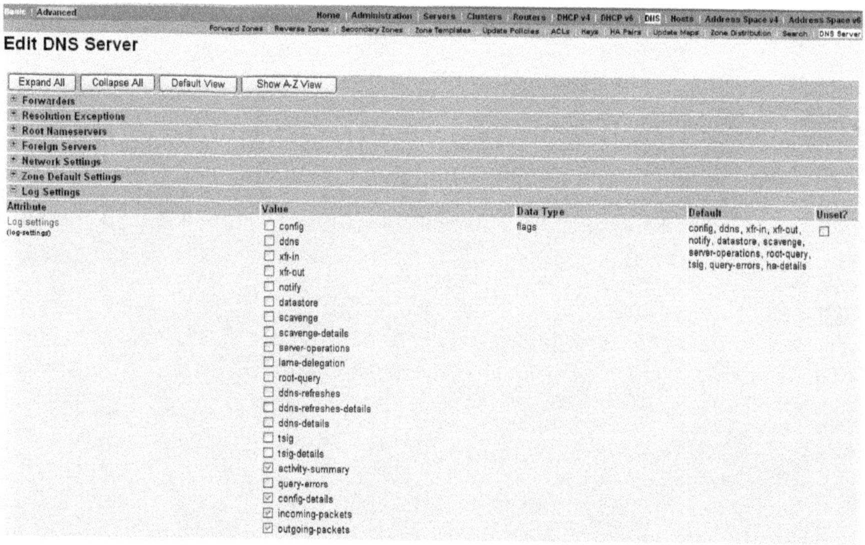

FIGURE 7.23   Enabling logging options on the CNR DNS server.

the resource records, the server should be able to respond to DNS queries for both IPv4 and IPv6.

7.2.3.1 Verifying DNS Server Operation To verify DNS server functionality you can try to perform a DNS lookup from the command prompt or try an ICMP ping using the host's fully qualified domain name (FQDN). The server should respond back from a lookup with all the addresses associated with that entry. An example of this is shown in Figure 7.24, where a query is performed on the **v6VideoServer.rcdn-cable.cisco.com** and the server replies with both IPv4 and IPv6 addresses associated with the entry. When the requesting client receives the reply from the server, it is up to the client or the application requesting this information to use either the IPv6 or the IPv4 address to establish the connection. For example, Windows Vista prefers IPv6 over IPv4. But if the application fails to establish a connection over IPv6, it can fall back to IPv4. This behavior may vary from one dual-stack implementation to another.

On the server side, the DNS server logs can display requests coming to the server and responses sent to requesting clients. Figure 7.25 illustrates the contents of the logs on the DNS server and displays the query requested by the client and the response returned by the server. The server replies with both the IPv4 and IPv6 addresses.

The DNS server operation can also be verified by using an ICMP ping from a host or a router. Figure 7.26 illustrates an ICMP ping on a router and the response received from the server.

```
bash-3.00# dig v6VideoServer.rcdn-cable.cisco.com any

; <<>> DiG 9.3.4-P1 <<>> v6VideoServer.rcdn-cable.cisco.com any
;; global options:  printcmd
;; Got answer:
;; ->>HEADER<<- opcode: QUERY, status: NOERROR, id: 1027
;; flags: qr aa rd ra; QUERY: 1, ANSWER: 2, AUTHORITY: 1, ADDITIONAL: 2

;; QUESTION SECTION:
;v6VideoServer.rcdn-cable.cisco.com. IN ANY

;; ANSWER SECTION:
v6VideoServer.rcdn-cable.cisco.com. 86400 IN AAAA 2001:420:a00:ffdb::253
v6VideoServer.rcdn-cable.cisco.com. 86400 IN A  10.89.240.253

;; AUTHORITY SECTION:
rcdn-cable.cisco.com.    86400   IN    NS     v6ProvServer.rcdn-cable.cisco.com.

;; ADDITIONAL SECTION:
v6ProvServer.rcdn-cable.cisco.com. 86400 IN A   10.89.240.254
v6ProvServer.rcdn-cable.cisco.com. 86400 IN AAAA 2001:420:a00:ffdb:203:baff:fe1f:bc68

;; Query time: 6 msec
;; SERVER: 10.89.240.254#53(10.89.240.254)
;; WHEN: Sun Sep 14 15:48:49 2008
;; MSG SIZE  rcvd: 187
```

FIGURE 7.24 Performing a DNS lookup on a hostname.

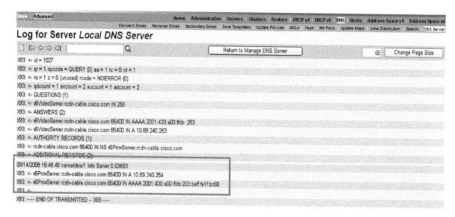

FIGURE 7.25   DNS server logs on CNR.

```
v6uBR7246VXR#ping v6ProvServer.rcdn-cable.cisco.com

Type escape sequence to abort.
Sending 5, 100-byte ICMP Echos to 2001:420:A00:FFDB:203:BAFF:FE1F:BC68, timeout is 2 seconds:
!!!!!
Success rate is 100 percent (5/5), round-trip min/avg/max = 0/0/4 ms
```

FIGURE 7.26   ICMP ping using FQDN of the IPv6 host.

Here the router is trying to ping the host using its FQDN. The router sends a DNS query to the server and receives a response with the IPv6 address of the host.

7.3   IPv6 SUPPORT ON TFTP SERVERS

IPv6 is supported on TFTP servers running certain operating system versions, such as Solaris version 9 or 10. It is also supported in the provisioning suite of Cisco broadband access center (BAC) version 4 and above. Only the Solaris implementation is discussed below.

7.3.1   Enabling TFTPv6 on Solaris 10

There is no current support for TFTPv6 in CNR, so we will look at how TFTPv6 can be enabled on Solaris 10. To enable this service on Solaris, you will need to issue the following command when logged in:

```
root@CNR# svcs|grep udp6
```

If there is no output for this command, the service definitions may be missing from the server. You will need to import the TFTPv6 service definitions from

```
<?xml version='1.0'?>
<!DOCTYPE service_bundle SYSTEM '/usr/share/lib/xml/dtd/service_bundle.dtd.1'>
<service_bundle type='manifest' name='export'>
  <service name='network/tftp/udp6' type='service' version='0'>
    <create_default_instance enabled='true'/>
    <restarter>
      <service_fmri value='svc:/network/inetd:default'/>
    </restarter>
    <exec_method name='inetd_start' type='method' exec='/usr/sbin/in.tftpd -s /tftpboot' timeout_seconds='0'>
      <method_context>
        <method_credential user='root' group='root'/>
      </method_context>
    </exec_method>
    <exec_method name='inetd_disable' type='method' exec=':kill' timeout_seconds='0'>
      <method_context/>
    </exec_method>
    <exec_method name='inetd_offline' type='method' exec=':kill_process' timeout_seconds='0'>
      <method_context/>
    </exec_method>
    <property_group name='inetconv' type='framework'>
      <propval name='converted' type='boolean' value='true'/>
      <propval name='source_line' type='astring' value='tftp dgram udp6 wait root /usr/sbin/in.tftpd in.tftpd -s /tftpboot'/>
      <propval name='version' type='integer' value='1'/>
    </property_group>
    <property_group name='inetd' type='framework'>
      <propval name='endpoint_type' type='astring' value='dgram'/>
      <propval name='isrpc' type='boolean' value='false'/>
      <propval name='name' type='astring' value='tftp'/>
      <propval name='proto' type='astring' value='udp6'/>
      <propval name='wait' type='boolean' value='true'/>
    </property_group>
    <stability value='External'/>
    <template>
      <common_name>
        <loctext xml:lang='c'>tftp</loctext>
      </common_name>
    </template>
  </service>
</service_bundle>
```

FIGURE 7.27   TFTPv6 service definitions.

a server already running TFTPv6 or and then try to reenable the service. To export the TFTPv6 service definitions, use the following command:

svccfg
svc:> export svc:/network/tftp/udp6

Which should generate output similar to that shown in Figure 7.27. Save the output of the command shown above to a file and give it a name like "TFTPv6," or any other suitable name. On the server you are trying to enable TFTPv6 to import the file by using the following command:

svccfg
svc:> import TFTPv6

The other option is to create the service definition by converting the /etc/inetd.conf entry into a service management facility (SMF) service manifest. Enable the tftp line in the /etc/inetd.conf file. Check that the entry reads as follows:

tftp dgram udp6 wait root /usr/sbin/in.tftpd in.tftpd -
s/tftpboot

This line prevents in.tftpd from retrieving any file other than the files that are located in /tftpboot. Convert the /etc/inetd.conf entry into a service management facility (SMF) service manifest:

 # /usr/sbin/inetconv

Make sure that the /tftpboot directory exists on the server and has the correct permissions for read/write access. Then enable the TFTPv6 service by using the following command:

svcadm enable svc:/network/tftp/udp6

Verify that service has been enabled on the server by using the **svcs** command. This time the output should be as follows:

svcs|grep udp6
online   Mar_04 svc:/network/tftp/udp6:default

7.3.2 Troubleshooting TFTPv6

To assure that the TFTPv6 service is working as desired, you can either view the log file on the server or use a packet capture tool to view the TFTPv6 message exchange between the client and the server. You can also verify TFTPv6 operation by transferring a file from the server. You will need to make sure that the file exists under the /tftpboot directory and has the correct read/write permissions. Figure 7.28 is an example of successful file transfer using TFTPv6.

Another way to troubleshoot TFTPv6-related issues is to use a packet capture tool to view the packet exchange between the client and the server (Figure 7.29). This example illustrates the contents of a TFTPv6 read request sent from a client (2001:DB8::2c81:c539:89fe:df) to the server (2001:DB8:1:2::3). Upon receiving this request, the server should respond back with the request data. If the request times-out, it could be due to several reasons:

- The server may not have received the request from the client, and the request could have been dropped by an intermediate device.
- The server may have received the request but due to being overloaded, was not able to respond, so the request may have been dropped by the server.
- The server may have received the request and responded, but the response may have been dropped by an intermediate device.

```
v6uBR7246VXR#copy tftp disk2:
Address or name of remote host []? 2001:BB:CAB:10:203:BAFF:FE1F:BC68
Source filename []? TFTPv6_SD.txt
Destination filename [TFTPv6_SD.txt]?
Accessing tftp://2001:BB:CAB:10:203:BAFF:FE1F:BC68/TFTPv6_SD.txt...
Loading TFTPv6_SD.txt from 2001:BB:CAB:10:203:BAFF:FE1F:BC68: !
[OK - 1753 bytes]

1753 bytes copied in 0.220 secs (7968 bytes/sec)
```

FIGURE 7.28 File transfer using TFTPv6.

```
Ethernet II, Src: Cisco_01:ae:4d (00:07:0e:01:ae:4d), Dst: 00:1d:35:f1:3b:55 (00:1d:35:f1:3b:55)
  Destination: 00:1d:35:f1:3b:55 (00:1d:35:f1:3b:55)
  Source: Cisco_01:ae:4d (00:07:0e:01:ae:4d)
  Type: IPv6 (0x86dd)
  Trailer: C575FDD5
Internet Protocol Version 6
  Version: 6
  Traffic class: 0x80
  Flowlabel: 0x00000
  Payload length: 28
  Next header: UDP (0x11)
  Hop limit: 64
  Source address: 2001:DB8::2c81:c539:89fe:df
  Destination address: 2001:DB8:1:2::3
User Datagram Protocol, Src Port: 51128 (51128), Dst Port: tftp (69)
  Source port: 51128 (51128)
  Destination port: tftp (69)
  Length: 28
  Checksum: 0xd6fd [correct]
Trivial File Transfer Protocol
  Opcode: Read Request (1)
  Source File: platinum.cm
  Type: octet
```

FIGURE 7.29   TFTPv6 packet capture.

If the TFTPv6 operation fails due to any of the reasons mentioned above or for some other reason, using the troubleshooting tips described in this section should help isolate the source of the problem and resolve it.

7.3.2.1 Verifying TFTPv6 Operation on Solaris 10   The TFTPv6 operation on a Solaris 10 server can be verified using several different techniques. One of them is use of the *svcs* command, as noted in Section 7.3.1. If the TFTPv6 service is enabled, the output should be similar to the following:

```
svcs|grep udp6
online   Jun_06 svc:/network/tftp/udp6:default
```

Another technique is to use the snoop utility to capture and view TFTPv6 packets exchanged between the client and the server, as shown in Figure 7.30, a such a TFTPv6 message exchange.

7.4   IPv6 SUPPORT ON AAA AND RADIUS SERVERS

Authentication, authorization, and accounting (AAA) (RFC2903) defines a framework used to provide network access and authorization to users connecting to network resources using various transport mechanisms. Users are typically authenticated and authorized using AAA either locally on the access

```
bash-3.00# /usr/sbin/snoop -d eri1 | grep TFTP
Using device /dev/eri (promiscuous mode)
2001:bb:cab:30::2 -> 2001:bb:cab:10:203:baff:fe1f:bc68 TFTP Read "TFTPv6_SD.txt" (octet)
2001:bb:cab:10:203:baff:fe1f:bc68 -> 2001:bb:cab:30::2 TFTP Data block 1 (512 bytes)
2001:bb:cab:30::2 -> 2001:bb:cab:10:203:baff:fe1f:bc68 TFTP Ack  block 1
2001:bb:cab:10:203:baff:fe1f:bc68 -> 2001:bb:cab:30::2 TFTP Data block 2 (512 bytes)
2001:bb:cab:30::2 -> 2001:bb:cab:10:203:baff:fe1f:bc68 TFTP Ack  block 2
2001:bb:cab:10:203:baff:fe1f:bc68 -> 2001:bb:cab:30::2 TFTP Data block 3 (512 bytes)
2001:bb:cab:30::2 -> 2001:bb:cab:10:203:baff:fe1f:bc68 TFTP Ack  block 3
2001:bb:cab:10:203:baff:fe1f:bc68 -> 2001:bb:cab:30::2 TFTP Data block 4 (217 bytes) (last block)
2001:bb:cab:30::2 -> 2001:bb:cab:10:203:baff:fe1f:bc68 TFTP Ack  block 4
```

FIGURE 7.30 Snooping TFTPv6 traffic on Solaris 10.

server (ER) or through an AAA server connected to the ER. Once users are authenticated and authorized, they are allowed access to network resources. The AAA framework can also be used to collect accounting data for users that are accessing network resources and services. The AAA framework uses a standard protocol such as remote authentication dial-in user service (RADIUS) (RFC2865) to carry AAA messages between an ER and an AAA server. The RADIUS protocol is widely used and implemented to manage access to network services. It defines a standard for information exchange between an ER and an AAA server for performing authentication, authorization, and accounting operations. A RADIUS AAA server can manage user profiles for authentication (verifying username and password) and provide configuration information specifying the type of service to deliver and enforce policies that can restrict user access.

To support IPv6 on RADIUS AAA servers, new attributes were defined to transport and store IPv6-related information. RFC 3612 specifies the operation of RADIUS over IPv6 as well as the RADIUS attributes used in support of IPv6 network access. These additional attributes are listed below.

- *NAS-IPv6-Address* This attribute identifies the IPv6 address of the ER that is requesting authentication of the user.
- *Framed-Interface-Id* This attribute identifies the IPv6 interface identifier to be configured for the user.
- *Framed-IPv6-Prefix* This attribute provides the IPv6 prefix and the corresponding route that needs to be configured for the user.
- *Login-IPv6-Host* This attribute indicates the system with which to connect the user when the Login-Service attribute is included.
- *Framed-IPv6-Route* This attribute provides routing information to be configured for the user on the ER.
- *Framed-IPv6-Pool* This attribute contains the name of an assigned pool that SHOULD be used to assign an IPv6 prefix for the user.

7.4.1 Generic AAA Configuration on an ER

To provide users with an IPv6 prefix and other network parameters, the ER needs to be configured with the relevant AAA and IPv6 configuration.

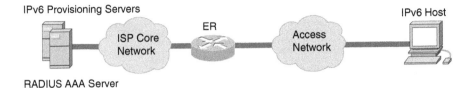

FIGURE 7.31 IPv6 provisioning using a RADIUS AAA server.

Figure 7.31 represents a typical scenario of how users connect to the ER and receive an IPv6 prefix from the RADIUS AAA server. The IPv6 host connects to the ER through the access network using any of the transport mechanisms discussed in Chapter 4. The ER is upgraded to dual-stack status to support both IPv4 and IPv6 connections. The ER is connected to the RADIUS AAA server over the ISP core network using any of the techniques described in Chapter 6. The RADIUS AAA server is upgraded to dual-stack mode to support both IPv4 and IPv6 clients and is RFC3162-compliant. The connection between the ER and the RADIUS AAA server could be IPv4 or IPv6. In other words, an IPv4 transport can also be used to carry IPv6 requests between the ER and the RADIUS AAA server.

When the user connects to the ER, it is authenticated and authorized by the RADIUS AAA server. Once the user passes authentication and authorization, it is assigned an IPv6 prefix and other network parameters. The IPv6 address assignment can be done by either using PPP (if the user connects to the ER over a PPP connection) or via DHCPv6. When using PPP for IPv6 address assignment, the user is authenticated and authorized by the RADIUS AAA server, but it receives the other network parameters from the DHCPv6 server using stateless DHCPv6. If DHCPv6 is used for IPv6 address assignment and to obtain other network parameters, the user is still authenticated and authorized by the RADIUS AAA server. The ER acts as a DHCPv6 relay agent and forwards DHCPv6 messages between the user and the DHCPv6 server.

Once IPv6 provisioning is complete, the user connects to the IPv6 network using the ER as its IPv6 default gateway. The ER installs a static route in its routing table for the IPv6 prefix assigned to the user. Let's a take a look at an example configuration on the ER and RADIUS AAA server for PPP-based connections.

7.4.1.1 IPv6 Configuration on the ER for a /64 Prefix Assignment (PPP Connection)
Next, we look at a typical IPv6 configuration on the ER running Cisco IOS for a PPP-based connection. In this example the user is assigned a /64 prefix from the RADIUS AAA server and uses SLAAC to assign itself a 128-bit IPv6 address.

```
ipv6 unicast-routing
! Enable IPv6 unicast routing in global configuration mode.
!
```

```
aaa new-model
aaa authentication ppp default group radius
aaa authorization network default group radius
! Generic AAA configuration on the ER for authentication and
authorization.
!
interface Virtual-Template 1
 ipv6 enable
 no ipv6 nd suppress-ra
! By default, RA messages are suppressed on point-to-point
links. This command enables RA messages to be sent out this
interface.
 ipv6 nd other-config-flag
! Set the O-bit in the RA messages to instruct the host to
use stateless DHCPv6 to obtain other network parameters
from the DHCPv6 server.
 ipv6 nd prefix framed-ipv6-prefix
! Advertise the IPv6 prefix obtained from the RADIUS server.
 ipv6 dhcp relay destination 2001:db8:1234:5678::1
! Configure DHCPv6 relay functionality to forward DHCPv6
messages to the DHCPv6 server.
 ppp authentication chap
! PPP authentication for the user.
!
radius-server host 10.89.240.22
radius-server key radius-password
! IP address of the RADIUS AAA server and the shared
password used to authenticate the ER connection.
```

7.4.1.2 IPv6 Configuration on the RADIUS AAA Server for a /64 Assignment (PPP Connection)

The following is the configuration on the RADIUS AAA server to assign a /64 IPv6 prefix to the user:

```
''IPv6-User1''
Auth-Type=Local, Password=''test''
User-Service-Type=Framed-User,
Framed-Protocol=PPP,
framed-ipv6-prefix=''ipv6:prefix#1=2001:db8:bad:f00d::
/64'',
framed-interface-id=''0:0:0:1'',
! User profile for user ''IPv6-User1'' on the RADIUS AAA
server.
```

The profile above is used for authenticating and assigning a /64 prefix to a user named "IPv6-User1" with the password "test." The IPv6 prefix to be assigned to this user is 2001:db8:bad:f00d::/64. Upon receiving this prefix, the user can assign itself an IPv6 address using SLAAC (since the A-bit is set to 1 in the RA messages sent by the ER). After assigning an IPv6 address, the user can use stateless DHCPv6 to obtain other network parameters (since the O-bit is set to 1 in the RA messages sent by the ER).

7.4.1.3   IPv6 Configuration on the ER for DHCP-PD Assignment (PPP Connection)
Let's take a look at an example configuration on the ER and RADIUS AAA server for a DHCP-PD assignment to a GWR connected to the ER as illustrated in Figure 7.32.

Here is a typical IPv6 configuration on an ER running Cisco IOS:

```
ipv6 unicast-routing
! Enable IPv6 unicast routing in global configuration mode.
!
aaa new-model
aaa authentication ppp default group radius
aaa authorization network default group radius
aaa authorization configuration IPv6LIST group radius
! Generic AAA configuration on the ER for authentication and
authorization.
!
interface
 ipv6 enable
 no ipv6 nd suppress-ra
! By default, RA messages are suppressed on point-to-point
links. This command enables RA messages to be sent out this
interface.
 ipv6 nd other-config-flag
! Set the O-bit to 1 in the RA messages to instruct the host
to use stateless DHCPv6 to obtain other network parameters
from the DHCPv6 server.
 ipv6 nd managed-config-flag
! Set the M-bit to 1 in the RA message to instruct the GWR to
use stateful DHCPv6 for address assignment.
 ipv6 nd prefix framed-ipv6-prefix
```

FIGURE 7.32   DHCP-PD assignment for GWR connected to the ER.

```
! Advertise the IPv6 prefix obtained from the RADIUS server.
 ipv6 dhcp relay destination 2001:db8:1234:5678::1
! Configure DHCPv6 relay functionality in order to forward
DHCPv6 messages to the DHCPv6 server.
 ipv6 dhcp server IPv6-Pool1
! IPv6 DHCP address pool.
 ppp authentication chap
! PPP authentication for the user.
 !
ipv6 dhcp pool IPv6-Pool1
 prefix-delegation aaa method-list IPv6LIST
! Define the pool to be used for DHCP-PD.
 !
radius-server host 10.89.240.22
radius-server key radius-password
! IP address of the RADIUS AAA server and the shared
password used to authenticate the ER connection.
```

In the configuration above, the GWR will first use DHCPv6 for IPv6 address assignment on its GWR-ER (upstream) link. Then it sends a DHCP-PD request for prefix delegation (the GWR needs to be configured as a DHCP-PD requesting router). Once it receives the DHCP-PD-delegated prefix, it will break the prefix into /64 chucks and send them out in the RA messages to the hosts connected to its downstream interfaces. The hosts use this /64 prefix to configure themselves with an IPv6 address.

IPv6 Configuration on the RADIUS AAA server for DHCP-PD Assignment (PPP Connection)   When the link between the GWR and ER is numbered with a global IPv6 address, two user profiles are stored on the RADIUS AAA server: "IPv6-User1" and "DHCPv6PD-User1." The "IPv6-User1" profile stores the GWR-ER link prefix, and "DHCPv6PD-User1" stores the /48 delegated prefix. These profile names are sent by the GWR during the authentication phase and need to match on the RADIUS AAA server.

```
``IPv6-User1''
Auth-Type=Local, Password=``test123''
User-Service-Type=Framed-User,
Framed-Protocol=PPP,
framed-ipv6-prefix=``ipv6:prefix#1=2001:db8:f00:b00::
/64''
framed-interface-id=``0:0:0:1'',
! User profile for the link between the GWR and ER. A /64
prefix is sent for this link.
``DHCPv6PD-User1''
Auth-Type=Local, Password=``test123''
```

```
User-Service-Type=Framed-User,
Framed-Protocol=PPP,
framed-ipv6-pool=''ipv6:prefix#1=2001:db8:f00d::/48''
! DHCP-PD user profile for the GWR, which configured as a
DHCP-PD requesting router.
```

When the link between a GWR and an ER is not numbered with a global IPv6 address (using a link-local IPv6 address), only one user profile is stored on the RADIUS AAA server: The "DHCPv6PD-User1" profile stores the /48 delegated prefix.

```
''DHCPv6PD-User1''
Auth-Type=Local, Password=''foo''
User-Service-Type=Framed-User,
Framed-Protocol=PPP,
framed-ipv6-pool=''ipv6:prefix#1=2001:db8:d00c::/48''
! User profile for DHCP-PD prefix assignment.
```

7.5 TROUBLESHOOTING IPv6 ON AN ER AND ON RADIUS AAA SERVERS

The troubleshooting methodology for IPv6 is similar to troubleshooting IPv4-related issues on the ER and on RADIUS AAA servers. You need to verify the configuration on the ER as well as on the RADIUS AAA server. Several show commands and debugs are available in Cisco IOS to troubleshooting IPv6-related issues.

7.5.1 Troubleshooting AAA and IPv6 Configurations on an ER

Several show commands and debugs are available to troubleshoot configuration-related issues on an ER. The following debugs and show commands can be used to troubleshoot AAA and IPv6 configuration-related problems:

debug aaa authentication. for troubleshooting authentication-related issues.

debug aaa authorization. for troubleshooting authorization-related issues.

debug aaa per-user. for troubleshooting per-user attributes.

debug radius. for generic RADIUS protocol debugging information.

debug radius authentication. for debugging RADIUS authentication packets.

show aaa servers. to show all AAA servers as seen by the AAA server MIB.

show aaa user. to display AAA users active in the AAA subsystem.

The following debugs can be used to troubleshoot IPv6-related problems on an ER:

debug ipv6 dhcp. for troubleshooting DHCPv6-related issues.

debug ipv6 nd. for troubleshooting IPv6 neighbor discovery messages.

show ipv6 dhcp binding. to view current DHCP binding for IPv6.

7.5.2 Troubleshooting User Profile and VSA Configurations on a RADIUS AAA Server

Apart from the show commands and debugs available on an ER to trouble-shoot AAA- and IPv6-related issues, you can also look at the logs on the RADIUS AAA server to troubleshoot IPv6 provisioning-related issues. An example follows:

```
*Oct 11 12:24:04.375: RADIUS/ENCODE: Best Local IP-Address
10.60.17.249 for Radius-Server 10.60.17.239
*Oct 11 12:24:04.375: RADIUS: Received from id 1646/29
10.60.17.239:1813, Accounting-response, len 20
*Oct 11 12:24:04.379: %LINK-3-UPDOWN: Interface Virtual-
Access3, changed state to down
*Oct 11 12:24:05.375: %LINEPROTO-5-UPDOWN: Line protocol
on Interface Virtual-Access3, changed state to down
*Oct 11 12:24:28.983: RADIUS/ENCODE(0000023D):Orig.
component type=VPDN
*Oct 11 12:24:28.983: RADIUS: AAA Unsupported Attr:
interface [169] 15
*Oct 11 12:24:28.983: RADIUS: 55 6E 69 71 2D 53 65 73 73 2D 49
44 35 [Uniq-Sess-ID5]
*Oct 11 12:24:28.983: RADIUS(0000023D): Config NAS IP:
0.0.0.0
*Oct 11 12:24:28.983: RADIUS/ENCODE(0000023D):
acct_session_id: 578
*Oct 11 12:24:28.983: RADIUS(0000023D): sending
*Oct 11 12:24:28.983: RADIUS/ENCODE: Best Local IP-Address
10.60.17.249 for Radius-Server 10.60.17.239
*Oct 11 12:24:28.983: RADIUS(0000023D): Send Access-
Request to 10.60.17.239:1812 id 1645/6, len 109
*Oct 11 12:24:28.983: RADIUS: authenticator 9E BC 6E 57 0E
F2 95 47-5A E9 C5 55 A8 E2 7D 13
*Oct 11 12:24:28.983: RADIUS: Framed-Protocol [7] 6 PPP [1]
*Oct 11 12:24:28.983: RADIUS: User-Name [1] 12
''u1@l2tp.de''
```

Oct 11 12:24:28.983: RADIUS: CHAP-Password [3] 19
! The username and password of the client are received by
the RADIUS server. Once the server authenticates the user,
it can assign an IPv6 prefix to the client.
*Oct 11 12:24:28.983: RADIUS: NAS-Port-Type [61] 6 Virtual
[5]
*Oct 11 12:24:28.983: RADIUS: NAS-Port [5] 6 570
*Oct 11 12:24:28.983: RADIUS: NAS-Port-Id [87] 17
''Uniq-Sess-ID570''
*Oct 11 12:24:28.983: RADIUS: Connect-Info [77] 11
''149760000''
*Oct 11 12:24:28.983: RADIUS: Service-Type [6] 6 Framed [2]
*Oct 11 12:24:28.983: RADIUS: NAS-IP-Address [4] 6
10.60.17.249
*Oct 11 12:24:28.987: RADIUS: Received from id 1645/6
10.60.17.239:1812, Access-Accept, len 67
*Oct 11 12:24:28.987: RADIUS: authenticator BA D3 EC FE 0A
57 CA 3D—AF 8D 2B C1 27 BD 5B 99
*Oct 11 12:24:28.987: RADIUS: Service-Type [6] 6 Framed [2]
*Oct 11 12:24:28.987: RADIUS: Framed-Protocol [7] 6 PPP [1]
*Oct 11 12:24:28.987: RADIUS: Vendor, Cisco [26] 35
*Oct 11 12:24:28.987: RADIUS: Cisco AVpair [1] 29
''ipv6:prefix#1=2003:1:1::/48''
*! IPv6 Prefix being assigned to the user **u1@l2tp.de**.*
*Oct 11 12:24:28.987: RADIUS(0000023D): Received from id
1645/6
*Oct 11 12:24:28.991: RADIUS/ENCODE(0000023D):Orig.
component type=VPDN
*Oct 11 12:24:28.991: RADIUS(0000023D): Config NAS IP:
0.0.0.0
*Oct 11 12:24:28.991: %LINK-3-UPDOWN: Interface Virtual-
Access3, changed state to up

7.6 SUMMARY

Enabling IPv6 on provisioning servers is an integral part of deploying IPv6 in
SP networks. It provides SP with the option to provision and manage a large
number of devices in their network and offer IPv6-based services to end
customers. It is important for the SP to evaluate their current provisioning
and management servers and verify if they can be upgraded to support IPv6.

The decision to upgrade a server to support IPv6 may depend on whether the
server needs to process and/or store any IPv6 information. For example, if a
management server is interfacing with a dual-stack device, it can poll IPv6-
related information using an IPv4 transport as well. But if the management

server is interfacing with an IPv6-only device, it would need to use IPv6 transport. SPs may decide to upgrade their existing IPv4 servers to support IPv6, or they may use different servers for IPv4 and IPv6. The key is for both services to work seamlessly without affecting the other.

For end customers it should be transparent whether an SP provides services over IPv4 or IPv6. Depending on the SP policy and the deployment model, end customers can receive anywhere from a /64 to a /48 IPv6 address assignment. With IPv4, end customers (residential users) typically receive a single /32 IPv4 address. As long as end customers are able to run their favorite applications and receive the services they desire from the SP, they should be able to take advantage of the benefits of IPv6 (permanent prefix assignment, transparent end-to-end communication, stateless address autoconfiguration, etc.) and its huge address space.

REFERENCES

1. C. Laat, G. Gross, L. Gommans, J. Vollbrecht, and D. Spence, "Generic AAA Architecture," RFC2903, August 2000.
2. C. Rigney, S. Willens, A. Rubens, and W. Simpson, "Remote Authentication Dial In User Service (RADIUS)," RFC2865, June 2000.
3. T. Narten, and R. Draves, "Privacy Extensions for Stateless Address Autoconfiguration in IPv6," RFC3041, January 2001.
4. B. Aboba, G. Zorn, and D. Mitton, "RADIUS and IPv6," RFC3162, August 2001.
5. R. Droms et al. "Dynamic Host Configuration Protocol for IPv6 (DHCPv6)," RFC3315, July 2003.
6. O. Troan, and R. Droms, "IPv6 Prefix Options for Dynamic Host Configuration Protocol (DHCP) Version 6," RFC3633, December 2003.
7. S. Thomson, C. Huitema, V. Ksinant, and M. Souissi, "DNS Extensions to Support IP Version 6," RFC3596, October 2003.
8. R. Droms, "Stateless Dynamic Host Configuration Protocol (DHCP) Service for IPv6," RFC3736, April 2004.
9. J. Brzozowski, K. Kinnear, B. Volz, and S. Zeng, "DHCPv6 Leasequery," RFC5007, September 2007.
10. Cisco Systems, Inc., "Release Notes for Network Registrar, 7.0." Available at http://www.cisco.com/en/US/partner/docs/net_mgmt/network_registrar/7.0/release/notes/CNR70ReleaseNotes.html.
11. Cisco Systems, Inc., "Cisco IOS IPv6 Configuration Library." Available at http://www.cisco.com/en/US/docs/ios/12_2t/ipv6/ipv6_c.html.

8 Conclusion

IPv6 has been deployed in several service provider (SP) core networks around the globe for a variety of reasons, such as long-term expansion plans and service-offering strategies, preparing for new service types, gaining competitive advantage, and so on. Recently, IPv6 is being actively deployed in service provider broadband access networks. In this book we highlight primary challenges that service providers face when deploying IPv6 in their broadband access networks. Therefore, after providing a brief overview of IPv6 business case and deployment models in broadband access networks, protocol basics, and provisioning in Chapters 1 and 2, we concentrated on IPv6 deployment techniques for broadband SP networks, such as cable, DSL, ETTH, and wireless. In Chapters 3 and 4 we compared IPv4 and IPv6 service deployment models and highlighted similarities and differences between the two. Subsequent chapters were focused on actual configuration of various network elements in delivering IPv6 broadband solutions. Guidelines for debugging and troubleshooting were also covered to help readers build more confidence in handling day-to-day operations involving provisioning, deployment and troubleshooting problems. In general, in this book we highlight various issues concerning IPv6 deployment in SP broadband access networks and suggest solutions and workarounds for them. Additional solutions are actually work in progress currently under discussion in the IETF and other standards bodies.

Due to the well-defined narrow focus of this book, we acknowledge that there are related interesting topics that would be of interest to our readers, yet are beyond our scope or receive limited coverage here:

IPv6 addressing considerations, IPv4-IPv6 and IPv6-IPv6 interworking, subscriber logging and recovery options. We would like to bring these topics to the attention of readers while noting that these issues are still evolving and are therefore moving targets. We offer some comments and notes on these topics below, although they deserve additional attention.

8.1 IPv6 ADDRESSING CONSIDERATIONS

Before an SP can deploy and scale IPv6 in its network, it needs to plan and design a related addressing scheme. A proper, well-thought-out, and scalable

Deploying IPv6 in Broadband Access Networks, By Adeel Ahmed and Salman Asadullah
Copyright © 2009 John Wiley & Sons, Inc.

addressing scheme is an integral part of IPv6 services deployment strategy in broadband SP networks. The SP needs to consider the following key points:

- Assignment of IPv6 addresses to infrastructure links
- What prefix length to use (/64, /126, /127, etc.) on point-to-point links
- When to use unique-local instead of global-unicast IPv6 addresses
- Whether end users should be allocated a /64, /56, or /48 IPv6 prefix

We have presented an overview of common IPv6 addressing strategies and suggested additional guidelines. However, it is strongly recommended that readers interested in deploying IPv6 keep abreast with IETF and regional Internet registry (RIR) activities for current best practices recommendations on the subject. There are few active IETF drafts, such as *draft-ietf-v6ops-addcon*, which discuss IPv6 addressing considerations, issues, workarounds, and related suggestions. We recommend that readers also follow up on he activities of the IFTF's 6man, v6ops, and shim6 working groups for the latest developments on this topic.

8.2   IPv4-IPv6 AND IPv6-IPv6 INTERWORKING

Although NAT-PT (RFC2766) has been moved to historical status (RFC4966) by the IETF, local network protection (RFC4864) covert ways in which IPv6 can provide similar functionality and benefits without using NAT. Deploying dual-stack networks where possible is the commonly used approach by SP to provide IPv4 and IPv6 services to customers. Time and experience with IPv6 deployment is proving that IPv4-IPv6 and IPv6-IPv6 interworking will be required for years to come. This realization has kicked-off several discussions and yielded IETF drafts that present new ideas and methodologies for IPv4-IPv6 and IPv6-IPv6 interworking, such as *draft-mrw-behave-nat66, draft-arkko-townsley-coexistence, draft-wing-nat-pt-replacement-comparison, draft-durand-softwire-dual-stack-lite, draft-bagnulo-behave-nat64, draft-bagnulo-behave-dns64, draft-baker-behave-v4v6-framework, and draft-baker-behave-v4v6-translation*. We recommend that readers follow up on these evolving discussions in the IFTF's v6ops, behave, and softwires working groups for the latest updates on this topic.

8.3   SUBSCRIBER LOGGING

IPv6 offers a number of autoconfiguration and privacy-extension options (RFC3041) for address assignment, but regulatory constraints make it challenging for an SP to benefit from these features. SPs are often required by law enforcement agencies to log the IP address used by a subscriber for tracking purposes if legal action is required against this particular subscriber. With IPv6, subscribers are typically assigned prefixes instead of being assigned a single IP address, which may make it difficult for the SP to associate an individual IPv6 address with a particular subscriber. The logging system would record the prefix allocated to a given subscriber site (GWR) and not a specific host address behind the GWR. If privacy extensions are used for generating the interface ID (the last 64 bits) of the IPv6 address (which changes frequently), it would make it difficult for the SP to track the IP addresses of end devices.

Some PPP-based deployment models allow an SP to assign a single IPv6 address to a subscriber, which can provide the subscriber with an address mapping correlation. This scenario is not always preferred, as a single IPv6 address assignment model would be difficult for the SP to manage and scale with thousands of subscribers connecting to its network. Subscriber logging becomes more challenging and less granular if the subscriber being tracked is located behind a NAT device, as the SP may not be able to track every end device in a subscriber's network. So a subscriber to IP address mapping may exist only for the outside interface of the NAT device or the IPv6 prefix assigned to the subscriber.

Although IPv6 SLAAC, along with privacy extensions, seems to be a legitimate privacy protection mechanism, it poses a problem for the SP in public hot spots, as it makes it more challenging to track subscribers properly, for the reasons mentioned above.

8.4 RECOVERY OPTIONS

The IPv4 address allocations are typically temporary, and only a single IP address is assigned to end devices or hosts. Due to this IPv4 address assignment model, a range of applications have been proven difficult and costly to deploy. For example, when servers are hosted behind a NAT device and require incoming traffic to reach them, static configuration has to be implemented on the NAT device to allow users on the Internet to connect to the server. This complexity can drive up the cost of deploying such a solution, due to support-related issues caused by misconfiguration and other operational issues.

The IPv6 address allocations are typically done in prefixes instead of in single address assignments, which favor deployment of business applications or always-on services. Static and large subscriber allocations enable deployment of applications and services with fixed and global addresses that are suitable for servers hosted by end users, which must be reached from the Internet. The absence of NAT with an associated application layer gateway (ALG) restores the end-to-end model and makes application deployment less dependent on third-party software.

Where as the SP network is naturally redundant through a mesh of routers, the SP router (ER) that connects subscribers to other networks has a great impact on the service offerings. Although the ER is well secured and redundant, it is possible that the ER will reboot entirely and lose network states, resulting in service disruption for an SP's subscribers. In a point-to-point deployment model, rebooting of an SP router (ER) can result in loss of PPP connections between the subscriber devices (GWRs) and the ER, causing loss of IP connectivity for all customers connected to the ER.

On the other hand, in the ISP-operated deployment model (discussed in Chapters 2 and 4), if DHCPv6 is used for address assignment and the BRAS reboots, the DHCPv6 client will not reconnect to the DHCPv6 server if its address lease time is still valid. This may lead to a loss of IP connectivity for the subscriber after the BRAS reboots since the BRAS would not have any information about the subscriber's IPv6 prefix (it initially learned the subscriber's IPv6 prefix by snooping DHCPv6 messages between the client and the server). It may drop any traffic intended for the subscriber, as it would not know where to send this traffic.

If the BRAS is acting as a DHCPv6 relay, it may lose the knowledge of all DHCPv6 clients upon a reboot. It can use the DHCPv6 lease-query mechanism to retrieve subscriber information from the DHCPv6 server. If the BRAS is acting as a DHCPv6 server, it must use persistent storage to restore its states in case of a reboot. A number of options are available to restore subscriber connectivity if the BRAS fails. Upon detecting a failure on the BRAS, the subscriber end can restart DHCPv6 message exchange so that the SP network devices can restore connectivity to the subscriber by refreshing their IPv6 information. Detection of such an event can be based on such features as bidirectional forwarding detection (BFD), which is a lightweight protocol and offers subsecond convergence, or the subscriber device can rely on IPv6 ND messages for failure detection.

8.5 SUMMARY

The challenges faced by SPs in deploying IPv6 in their broadband access networks may vary based on the deployment model and SP customer business requirements. With proper planning and by following the design guidelines discussed in this book, SPs should be able to deploy and scale IPv6 successfully in their broadband environments and offer value-added services to their customers.

APPENDIX A
IPv6 Case Study

CONTRIBUTED BY ALEXANDRE CASSEN

A.1.1 Context and IPv6 Statement

Free is the second-largest French ISP with more than 4 million broadband subscribers (ADSL and FTTH). Free provide access to its network and services through its set-top box (STB) called *Freebox*, the name of Free's R&D lab in charge of hardware and software design. The Freebox STB has been designed entirely, from hardware to software, by the R&D team to provide flexibility and reactivity to introduce new features and services.

As an ISP, we wanted to find a transparent and rapid way to provide IPv6 to our customers. We wanted to offer IPv6 access to end users though a simple activation approach rather than through migration, so we investigated all available IPv6 network designs. The 6to4 approach had been evaluated but not considered to be a potential candidate because of a number of well-known issues (e.g., asymmetric routings, DoS exposition, not a global solution). We were evaluating several options when Rémi Després came up with the 6rd (IPv6 rapid development) design in November 2007. It was simple and straightforward enough that we could begin immediately to work on implementation. (For more information on 6rd, refer to Despris' IETF draft : draft-despres-6rd-02.txt.)

IPv6 provides designs to simplify and speed up future Internet applications, but these benefits may not justify the extra expense associated with deploying IPv6. On the one hand, most ISPs are waiting for "killer apps" that will provide business justification for mass investment in IPv6 deployment. On the other hand, applications vendors are delaying IPv6 application development due to the lack of IPv6 demand from customers. This holdup on IPv6 deployment was unlocked with 6rd design simplicity in our environment and enabled us easily to tweak and update CPE (Customer provider equipment) routing. Most current operating systems are dual-stacked, so if an ISP can provide both IPv4 and IPv6, it will be ready to offer its customers rapid, transparent traffic shift from IPv4 to IPv6 and thus properly address IPv4 exhaustion.

Deploying IPv6 in Broadband Access Networks, By Adeel Ahmed and Salman Asadullah
Copyright © 2009 John Wiley & Sons, Inc.

Using the 6rd approach, native IPv6 connection has been provided to our customers with no extra charge. Our deployment timeline was as follows:

- November 7, 2007: Despris presented 6rd and we decided to make IPv6 happen.
- November 9, 2007: We got an IPv6 prefix from RIPE.
- November 10, 2007: The first prototype of a 6rd gateway and 6rd support was ready for our CPE.
- December 11, 2007: Opt-in was made available to all our customers.
- March 2008: We began our first IPv6-only service launch: Telesite.

A.1.2 Free Network Environment

A free core network backbone is built around a set of CISCO CRS-1's. Those routers are federation points for every PoP within the network. For DGE routing, Free runs CISCO Cat6500 in order to concentrate every PoP aggregation to the core network. As a broadband ISP, Free provides ADSL and FTTH connections to its customers. ADSL access is provided by Freebox DSLAM and FTTH by CISCO Cat4500. The global network architecture is shown in Figure A.1.

Since some of the network equipment was IPv6 ready, we needed to find a solution for our DSLAM that did not support IPv6. Our goal was to find a solution to push IPv6 for every customer, FTTH and ADSL, in the same time frame rather than using a phased migration plan. The 6rd approach

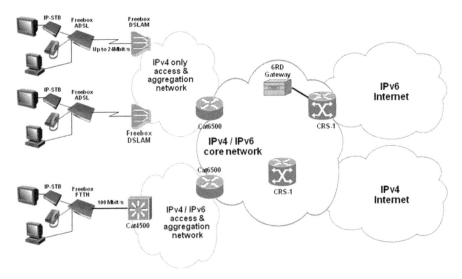

FIGURE A.1

offered a solution that fit our environment. We represented only a single 6rd gateway in Figure A.1, but for redundancy and traffic load balancing, multiple 6rd gateways have been installed in the core network. Those gateways are regular x86 PCs connected directly to CRS-1's via multiple 10-gigabit interfaces.

A.1.3 6rd Idea

6rd is built on 6to4 mechanisms, as described in RFC3056, to enable a service provider to deploy IPv6 unicast service rapidly to IPv4 sites to which it provides CPE. Like 6to4, it utilizes stateless IPv6 in IPv4 encapsulation to transit IPv4-only network infrastructure. Unlike 6to4, a 6rd service provider uses an IPv6 prefix of its own in place of the fixed 6to4 prefix (2002::/16). The purpose of 6rd is to modify 6to4 slightly so that:

- Packets coming from the global IPv6 Internet that enter 6rd gateways of an ISP are only packets destined to customer sites of this ISP.
- All IPv6 packets destined to 6rd customer sites of an ISP, and coming from anywhere else on the IPv6 Internet, traverse the 6rd gateway of this ISP.

To achieve those two points, 6rd specifies an IPv6 address format in a way to offer the benefits and simplicity of the regular 6to4 approach. In addition, 6rd addressing solves the well-known asymmetric routing path issue of 6to4. Steps in providing 6rd are:

1. Get an Internet IPv6 prefix from your RIR.
2. Install some 6rd gateways in the core backbone network.
3. Upgrade the CPE routing code to support 6rd encapsulation.

A.1.4 IPv6 Prefix and Addressing

An IPv6 address is built around 128 bits according to the specification shown in Figure A.2. 6rd proposes a specification for the 64-bit link prefix field. The key element is to embed a CPE IPv4 address inside this field, as shown in Figure A.3. The IPv6 prefix *mUST* be at most a 32-bit prefix in order to keep

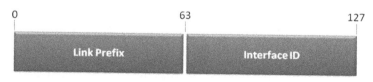

FIGURE A.2

FIGURE A.3

FIGURE A.4

the other 32 bits for a CPE IPv4 address inside, 64 bits of the link prefix. The subnet ID length will then depend on the IPv6 prefix length.

In our deployment use case we obtained from our RIR (RIPE) the /26 prefix of 2a01:0e00::/26. This prefix is then offered a 6 bits of free space to address the subnet at the customer's site. For simplicity and maintainability, we decided to reserve site prefix bits 27 and 28 to define our addressing policy:

- 0 is reserved the network administration.
- 1 and 2 are reserved for future use and deployment.
- 3 is reserved for 6rd.

In our deployment, a 6rd site address looks as shown in Figure A.4.

We can assign 2a01:0e30::/28 as the 6rd prefix. Subnet ID is used to offer a multiple IPv6 subnet inside the same sites. Subnet ID of 0 can be reserved to define the CPE administration link prefix, announced via the neighbor discover protocol.

A.1.5 6rd in Action

6rd operations can be summed up as shown in Figure A.5. To illustrate 6rd operation workflow, we describe three different use cases. The first is traffic originated from a customer site and directed to the IPv6 Internet, the second is traffic originated from a customer site and directed to another customer site, and the last is traffic coming from an IPv6 Internet host and targeting a customer site. As in 6to4, 6rd use an IPv4 anycast address for a 6rd gateway.

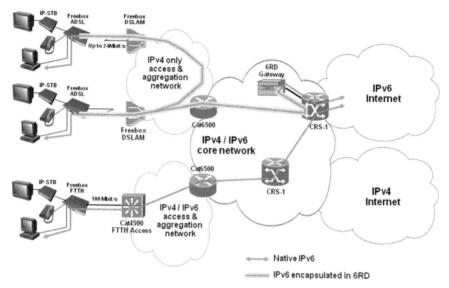

FIGURE A.5

We took a 192.88.99.k address, routed with the same /24 prefix as 6to4, but with a k other than 1, to avoid any confusion with the 6to4 address of RFC3068.

The first scenario occurs when a host of a customer site and behind the Freebox CPE (STB) tries to reach an IPv6 Internet host. For example, a computer behind a Freebox is trying to reach www.google.com. The computer is running native IPv6, so it resolves www.google.com to find the AAAA entry and start sending an IPv6 request. The CPE catches this IPv6 traffic and looks for a match to the global 6rd prefix (2a01:0e30::/28). Since www.google.com is outside our 6rd prefix, the CPE encapsulates traffic into IPv4 with the IPv4 source address of the CPE IPv4 router address and the IPv4 destination address of the 6rd gateway IPv4 anycast address. This IPv4 encapsulate traffic transit until the 6rd gateway is decapsulated and routed directly to the IPv6 Internet.

The second scenario occurs when an IPv6 Internet host is trying to reach a host at a customer site inside the 6rd prefix. CRS-1 is our IPv6 border router, its routing setup is simple. On the one hand, it announces its IPv6 prefix of 2a01:0e00::/26 to the IPv6 Internet; on the other hand, it defines and redistributes to the IPv6 Internet static routes to the 6rd gateway for the 6rd prefix of 2a01:0e30::/28. This last routing setting will route traffic coming from the IPv6 Internet and match the 6rd prefix to the 6rd gateway. At its IPv6 ingress, the 6rd gateway and the CPE are performing the same encapsulation operations. Thanks to 6rd addressing policy, a remote customer site is deduced by extracting the CPE IPv4 address from the IPv6 destination address. By now, the 6rd gateway can perform a stateless IPv4 operation and send this IPv4 encapsulated traffic to a remote customer site as in 6to4.

The third scenario involves traffic between customer sites. By restoring direct host access by removing NAT, IPv6 offers a nice design for any peer-to-peer application. Demand for such an application is growing; thus it was mandatory in our network design not to route all the intrasite IPv6 traffic to the 6rd gateway. To offer an answer to this point, as 6to4 does, 6rd at the CPE side routes traffic between customer sites based on 6rd prefix matching. The CPE of one customer site will route encapsulated IPv6 traffic directly to another customer site inside the same 6rd prefix.

A.1.6   Operational Considerations: At CPE Side

The Freebox CPE runs a Linux kernel, so it was straightforward to add support to 6rd since low-level software design for 6to4 is already part of the mainline code. Our Freebox CPE is then running two pieces of software in order to provide IPv6 connectivity to our customers. The first component is a router advertisement (RA) protocol stack, used to make our CPE act as an IPv6 router. This protocol stack is specified by RFC2461 and relies on sending RA messages periodically to the CPE local Ethernet LAN. It also replies to any router solicitation (RS) messages that it may receive from the CPE local Ethernet LAN. This protocol stack provides a simple way to provide for IPv6 stateless address autoconfiguration (SLAAC). Linux is implementing an IPv6 neighbor discover protocol called RADVD, whose configuration on the CPE side looks as follows:

```
interface br0 {
    IgnoreIfMissing off;
    AdvSendAdvert on;
    UnicastOnly off;
    AdvManagedFlag off;
    AdvOtherConfigFlag off;

    AdvLinkMTU 1480;

    prefix 2a01:e3x:xxxx:xxx0::/64 {
        AdvOnLink on;
        AdvAutonomous on;
        AdvValidLifetime 86400;
        AdvPreferredLifetime 86400;
    };

    RDNSS 2a01:e00::2 2a01:e00::1 {
    };
};
```

In our design we advertise the /64 prefix with subnet ID 0 to every node on the CPE local Ethernet LAN. One important RADVD configuration element is the AdvLinkMTU, which in our 6rd use case must take care of IPv4 encapsulation overhead. The next configuration element is the IPv6 routing table on the CPE side. This routing table is set by the following commands:

```
# ip tunnel add sit2 mode sit local xx.xx.xx.xx \
         6rd_prefix 2a01:e30::/28 ttl 64
# ip link set dev sit2 up
# ip -6 addr add 2a01:e3x:xxxx:xxx0::1/128 dev sit2
# ip -6 addr add 2a01:e3x:xxxx:xxx0::1/64 dev br0
# ip -6 route add default via ::192.88.99.201 dev sit2
metric 1
```

The configurations line above set a Linux SIT tunnel based on 6rd prefix matching and set a default IPv6 route to a 6rd anycast address. We intentionally obfuscated the CPE IPv4 address here by "xx.xx.xx.xx," so you should imagine replacing this value by a CPE Internet IPv4 address.

Linux natively supports a IPv6-over-IPv4 tunnel by implementing the simple Internet transition device driver. For our needs we tweaked the SIT driver to make it support 6rd encapsulation. We simply implemented "6rd_prefix" as a new tunneling helper. The Linux kernel patch for adding support to 6rd looks as follows:

```
— sit.c.orig      2009-02-09 14:21:39.000000000 +0100
+++ sit.c         2009-02-09 15:27:38.000000000 +0100
@@ -161,7 +161,9 @@
            h ^= HASH(local);
      }
      for (tp = &tunnels[prio][h]; (t = *tp) != NULL; tp =
&t->next){
-            if (local == t->parms.iph.saddr && remote ==
t->parms.iph.daddr)
+            if ((local == t->parms.iph.saddr && remote ==
t->parms.iph.daddr) &&
+                (!memcmp(&parms->6rd_zone, &t-
>parms.6rd_zone,
+                    sizeof (parms->6rd_zone)) &&
+                parms->6rd_prefix == t->parms.6rd_prefix))
+                return t;
      }
      if (!create)
@@ -416,6 +418,27 @@
      return dst;
 }
+static inline __be32 try_6rd(struct in6_addr *6rd_zone,
                u8 6rd_prefix, struct in6_addr *v6dst)
```

```
+{
+       __be32 dst = 0;
+
+       /* isolate zone according to mask */
+       if (ipv6_prefix_equal(v6dst, 6rd_zone, 6rd_prefix)) {
+               unsigned int d32_off, bits;
+
+               d32_off = 6rd_prefix >> 5;
+               bits = (6rd_prefix & 0x1f);
+
+               dst = (ntohl(v6dst->s6_addr32[d32_off])
+               if (bits)
+                   dst |= ntohl(v6dst->s6_addr32[d32_off + 1]) >>
+                       (32 - bits);
+               dst = htonl(dst);
+       }
+       return dst;
+}
+
/*
*+      This function assumes it is being called from
dev_queue_xmit()
*       and that skb is filled properly by that function.
@@ -445,7 +468,11 @@
        if (skb->protocol != htons(ETH_P_IPV6))
                goto tx_error;
-       if (!dst)
+       if (!dst && tunnel->parms.6rd_prefix)
+           dst = try_6rd(&tunnel->parms.6rd_zone,
+                       tunnel->parms.6rd_prefix,
+                       &iph6->daddr);
+       else if (!dst)
+           dst = try_6to4(&iph6->daddr);
        if (!dst) {
@@ -632,6 +659,11 @@
            if (p.iph.ttl)
                p.iph.frag_off |= htons(IP_DF);
+           /* prefix must be smaller than 95 bits since we
fetch
+           * an ip address after them */
+           if (p.6rd_prefix >= 95)
+               goto done;
+
+           t = ipip6_tunnel_locate(&p, cmd ==
SIOCADDTUNNEL);
```

```
                    if (dev != ipip6_fb_tunnel_dev && cmd ==
SIOCCHGTUNNEL) {
@@ -650,6 +682,8 @@
                             ipip6_tunnel_unlink(t);
                             t->parms.iph.saddr = p.iph.saddr;
                             t->parms.iph.daddr = p.iph.daddr;
+                            t->parms.6rd_zone = p.6rd_zone;
+                            t->parms.6rd_prefix = p.6rd_prefix;
                             memcpy(dev->dev_addr, &p.iph.saddr, 4);
                             memcpy(dev->broadcast, &p.iph.daddr,
4);
                             ipip6_tunnel_link(t);
```

A.1.7   Operational Considerations: At the 6rd Gateway Side

The 6rd gateway is, on the one hand, configuring 6rd encapsulation and, on the other, configuring routings to connect to a CRS-1. Figure A.6 illustrates a 6rd gateway interconnection with a CRS-1. The 6rd gateway is connected to the CRS-1 via two 10-gigabit interfaces. One 10-gigabit interface is dedicated for the IPv4 Internet and the other for the IPv6 Internet. The 6rd configuration looks as follows:

```
# ip addr add 212.27.58.6/30 dev eth2
# ip route add default via 212.27.58.5 dev eth2
# ip link set dev eth2 up

# ip -6 addr add 2a01:e00:1:2::2/126 dev eth3
# ip -6 route add default via 2a01:e00:1:2::1 dev eth3
# ip link set dev eth3 up

# ip addr add 192.88.99.201/32 dev dummy0
# ip link set dev dummy0 up

# ip tunnel add sit1 mode sit local 192.88.99.201 6rd_prefix
2a01:e30::/28 ttl 64
# ip -6 addr add 2a01:e00:1:2::2/126 dev sit1
# ip -6 route add 2a01:0e30::/28 via 2a01:e00:1:2::1 dev
sit1
# ip link set dev sit1 up
```

In this configuration we create an eth2 interface for the IPv4 world and an eth3 interface for the IPv6 world, with a default route for the IPv4 and IPv6 interfaces to the CRS-1. We create a dummy interface, which is similar to a loopback interface, and set a 6rd IPv4 anycast address. Finally, we define a 6rd sit interface to match our 6rd prefix and route traffic accordingly.

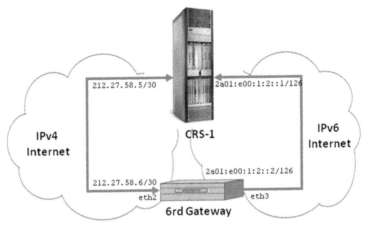

FIGURE A.6

A.1.8 Operational Considerations: At the CRS-1 Side

The CRS-1 configuration is similar to the 6rd configuration as to interface configuration. In addition, it configures an IPv6 BGP announcement in order to announce our IPv6 prefix to a remote peer. The interface configuration looks as follows:

```
interface TenGigE1/1/0/5
 description 6rd eth2
 ipv4 address 212.27.58.5 255.255.255.252
!
interface TenGigE1/1/0/7
 description 6rd eth3
 ipv6 address 2a01:e00:1:2::1/126
!
```

The CRS-1 routings are defined as follows:

```
router static
 address-family ipv4 unicast
  192.88.99.201/32 TenGigE1/1/0/5 212.27.58.6 tag 5002
 address-family ipv6 unicast
  2a01:e00::/26 Null0
  2a01:e30::/28 TenGigE1/1/0/7 2a01:e00:1:2::2 tag 5002
!
route-policy static_to_bgp
  if tag eq 5002 then
    set community (12322:65350, 12322:65400, 12322:64984)
```

```
   pass
  endif
end-policy
!
bgp confederation identifier 12322
 bgp router-id yy.yy.yy.yy
 bgp cluster-id zz.zz.zz.zz
 bgp graceful-restart
 bgp as-path-loopcheck
 bgp bestpath med always
 bgp bestpath med confed
 address-family ipv4 unicast
. . . .
  redistribute connected route-policy connected_to_bgp
  redistribute static route-policy static_to_bgp
 address-family ipv6 unicast
 network ::/0 route-policy default_to_bgp
 network 2a01:e00::/26 route-policy network_to_bgp
  redistribute connected route-policy connected_to_bgp
  redistribute static route-policy static_to_bgp
  !
```

We defined a static route for our 6rd prefix 2a01:0e30::/28 and IPv4 anycast address to our 6rd gateway. Those routes are redistributed inside our eBGP so that we announce our 6rd prefix to the IPv6 Internet.

A.1.9 That's It!

Using the 6rd approach we have been able to offer IPv6 connections to all our customers. This design provides an important flexibility to deploy native IPv6 at an edge router and concentration network by offering a transparent transition mechanism to the end user.

APPENDIX B
DHCPv6 Message Types
and Option Codes

DHCP MESSAGE TYPES

Registry Value	Description	Reference
0	Reserved	RFC3315
1	SOLICIT	RFC3315
2	ADVERTISE	RFC3315
3	REQUEST	RFC3315
4	CONFIRM	RFC3315
5	RENEW	RFC3315
6	REBIND	RFC3315
7	REPLY	RFC3315
8	RELEASE	RFC3315
9	DECLINE	RFC3315
10	RECONFIGURE	RFC3315
11	INFORMATION-REQUEST	RFC3315
12	RELAY-FORW	RFC3315
13	RELAY-REPL	RFC3315
14	LEASEQUERY	RFC3315
15	LEASEQUERY-REPLY	RFC5007
16–255	UNASSIGNED	RFC5007

Source: RFC3315

DHCP OPTION CODES

Registry Value	Description	Reference
0	Reserved	RFC3315
1	OPTION_CLIENTED	RFC3315

(Continued)

Deploying IPv6 in Broadband Access Networks, By Adeel Ahmed and Salman Asadullah
Copyright © 2009 John Wiley & Sons, Inc.

DHCP OPTION CODES (Continued)

Registry Value	Description	Reference
2	OPTION_SERVERID	RFC3315
3	OPTION_IA_NA	RFC3315
4	OPTION_IA_TA	RFC3315
5	OPTION_IAADDR	RFC3315
6	OPTION_ORO	RFC3315
7	OPTION_PREFERENCE	RFC3315
8	OPTION_ELAPSED_TIME	RFC3315
9	OPTION_RELAY_MSG	RFC3315
10	Unassigned	RFC3315
11	OPTION_AUTH	RFC3315
12	OPTION_UNICAST	RFC3315
13	OPTION_STATUS_CODE	RFC3315
14	OPTION_RAPID_COMMIT	RFC3315
15	OPTION_USER_CLASS	RFC3315
16	OPTION_VENDOR_CLASS	RFC3315
17	OPTION_VENDOR_OPTS	RFC3315
18	OPTION_INTERFACE_ID	RFC3315
19	OPTION_RECONF_MSG	RFC3315
20	OPTION_RECONF_ACCEPT	RFC3319
21	SIP Servers Domain Name List	RFC3319
22	SIP Servers IPv6 Address List	RFC3646
23	DNS Recursive Name Server Option	RFC3646
24	Domain Search List option	RFC3633
25	OPTION_IA_PD	RFC3633
26	OPTION_IAPREFIX	RFC3898
27	OPTION_NIS_SERVERS	RFC3898
28	OPTION_NISP_SERVERS	RFC3898
29	OPTION_NIS_DOMAIN_NAME	RFC3898
30	OPTION_NISP_DOMAIN_NAME	RFC4075
31	OPTION_SNTP_SERVERS	RFC4242
32	OPTION_INFORMATION_REFRESH_TIME	RFC4280
33	OPTION_BCMCS_SERVER_D	RFC4280
34	OPTION_BCMCS_SERVER_A	RFC4776
35	Unassigned	RFC4649
36	OPTION_GEOCONF_CIVIC	RFC4580
37	OPTION_REMOTE_ID	RFC4704
38	OPTION_SUBSCRIBER_ID	RFC5192
39	OPTION_CLIENT_FQDN	RFC4833
40	OPTION_PANA_AGENT	RFC4833
41	OPTION_NEW_POSIX_TIMEZONE	RFC4994

DHCP OPTION CODES (Continued)

Registry Value	Description	Reference
42	OPTION_NEW_TZDB_TIMEZONE	RFC5007
43	OPTION_ERO	RFC5007
44	OPTION_LQ_QUERY	RFC5007
45	OPTION_CLIENT_DATA	RFC5007
46	OPTION_CLT_TIME	RFC5007
47	OPTION_RELAY_DATA	RFC-ietf-mip6-hiopt-17.txt
48	OPTION_CLIENT_LINK	RFC-ietf-mip6-hiopt-17.txt
49	OPTION_MIP6_HNINF	RFC5223
50	OPTION_MIP6_RELAY	RFC-ietf-capwap-dhc-ac-option-02.txt
51	OPTION_V6LOST	
52	OPTION_CAPWAP_AC_V6	
53–255	Unassigned	

Source: RFC3315.

INDEX

Deploying IPv6 in Broadband Access Networks, By Adeel Ahmed and Salman Asadullah
Copyright © 2009 John Wiley & Sons, Inc.